Fake Science

Science Without Basis

By Steve Preston

2nd Edition

© Copyright 2020, Steve Preston
All rights reserved.

Table of Contents

FAKE SCIENCE 1
TABLE OF CONTENTS 4
INTRODUCTION 6
FAKE TREPANATION 10
HEROINE MEDICINE 15
MIRACLE OF RADIUM 20
FAKE BIOLOGY 27
FAKE ARYANISM 30
SCIENTIFIC POLYGENISM 34
CONSENSUS MONOGENISM 37
FAKE PHRENOLOGY 41
FAKE MOLEOSCOPY 46
FAKE PHYSIOGNOMY 52
FAKE SCIENTIFIC HISTORIES 56
LIGHT BULB INVENTION 60
TELEPHONE INVENTION 65
BATTERY INVENTION 68
NUCLEAR INVENTION 71
MOVIE MISUNDERSTANDING 77
HIDDEN FLYING 83
FAKE SCIENCE QUIPS 99
FAKE CLIMATOLOGY 102
ERROR OF NUCLEAR TIMERS 120
EARTH STABILITY FALLACY 132
FAKE MYSTERY OCEAN 136

GIANT REJECTION..**142**
PLEISTOCENE DINOSAUR ERROR..**148**
HIDDEN DRYAS...**155**
MISTAKEN CRUSADES..**173**
CONCLUSIONS..**185**
ABOUT THE AUTHOR..**188**

Introduction

This book is about one of the worst evils developed by mankind. I'm not talking about mass destruction, or war, or developing man-made disease or any of that. I'm simply talking about using the name 'scientist' to promote a cause that no scientist can disallow without horrible retribution. This establishes a masterful method to push any agenda with the backing of the scientific community, frozen from dissention. While the method is a common practice of governments in the past, it seems to be gaining in acceptance so we need to at least understand how this fake science works and where it has been used so we can assist in the limiting of its tragic results.

In this book we will go through a number of the Fake and Consensus-established scientific assertions made without valid research simply by stating a claim. To do this, the community governing the claim eliminates scientist-dissenters and establishes consensus by overpowering science, intimidating alternate assertions, glamorizing the consensus truth model, and getting the government behind the claims. Usually this ended and still ends in a level of disaster for everyone except the fake-scientist. Without a doubt, consensus has changed the makeup of discovery for all time. No need to use logic, science, or physical evidence when you have consensus of the elite scientists. It's hard to even believe how rampant this illogical, messy, and scary tactic was and has become. The fallout stretches from Medicine, Physics, Meteorology, Physiology, Astronomy, Engineering, and just about all sciences and many times we never even find out about it. Once the "error" is found out, text books still continue to keep the consensus lies going for many years as governments, and large communities of powerful entities do not want to disrupt the "truth" they had built.

The one thing that seems to be a constant in this type of fake science is that the initiators believe they are saving the world from the world. To do this, the initiators come to a conclusion and look for a cause to support the agenda. Let me give you a few examples.

Greek Teeth-In Greece it was scientifically decided that women were of a lower status than men so that is why they had fewer teeth. The strange thing about this nonsense is that it was never challenged for hundreds of years.

Train Speed- In America fake science determined that if a train went faster than 21 miles per hour the people on the train would not be able to breathe so no attempts were made to speed up the trains until the middle of the 19th century.

Bode's Law- consensus scientists kept teaching the absurd Bode's Law of planetary positioning even when it never approximated the positions of the planets. Oddly, it is still taught in schools.

No Rocks in Space- Until the 19th century, there was consensus by astronomers that no rocks were in space, so no rocks could fall from space. The strange part of this one is that many had witnessed meteorites falling from the sky but the science community of the 18th century would not accept this abomination.

Fake Anthropology- In Anthropology we find almost unbelievable sciences of Phenology, Aryanism, Craniometry, Polygenism, Moleoscopy, and Physiognomy. By using these fake-sciences, the world was to be saved by improving the development of man by inspection and sometimes eradication.

Consensus Engineering- In invention we find the reestablished developments of movies, telephones, radio, stoves, batteries, airplanes, and many other works. My using a consensus of development rather than the actual developments, the world was to be saved from looking at those who would not make America and Americans as great, I suppose.

Fake and Harmful Environmental Sciences- Here we find something called climatology. By using consensus-science the world was to be saved from people even when real science keeps proving the errors, miscalculations, and manipulation of historical records to support the "green Industries" by destroying many others. When it comes to a cause, money certainly reared its ugly head.

Consensus Astronomy- Besides Bode's Law and the idea that no rocks were in space, in astronomy we find consensus Geocentricity. People were even excommunicated for not believing what was spouted as truth to protect the world form blasphemy.

Consensus Physics- Besides describing the maximum train speed, we find a crazy adherence to nuclear decay timing after it was totally debunked. This saves the world from wondering about evolution and keeps the "good-half" of the historical scientists from looking bad which might make the ignorant masses start thinking on their own. Additionally, the possibility of the very early use of nuclear weaponry would make history books have to be rewritten which would drive confusion to the masses.

Consensus Earth Sciences- Here we find that the instability of the Earth has been hidden to keep fear away from our populations even if it makes many things not make sense at all.

Consensus Evolution-Here we find that the secret non-fossilize dinosaur finds and the huge pile of evidence about advanced civilization of ancient, giant people which are not discussed so people will be saved from knowing dinosaurs roamed the earth thousand years ago rather than hundreds of millions and that the concept of uncontrolled evolution doesn't work so people won't be drawn to the Creator God.

Fake Biology- Here we find that DNA mutation evidence eliminates the concept of "Out of Africa" development of mankind as a working model. This would possibly make people believe Biblical history and confusion would be put in our secular schools. The consensus scientist thinks to himself, "This cannot happen."

Consensus Medicine-In medicine we find something called Trepanation and heroine medication. These curative measures were developed to save the world from all disease including mental aberration. Speaking of mental aberration, let's look at trepanation.

.

Fake Trepanation

One of the many examples of Consensus Science over scientific research comes from the area of medicine. We will be looking at a variety of these medical procedures performed by consensus rather than having a documented reason tested by science of any kind. One that comes to mind is a practice that has been performed for over 30 thousand years. Neanderthal skulls have been found that have had a hole drilled into the skull for so medical relief. Like the other elements discussed in this book, trepanation was practiced worldwide and accepted as proper medical treatment for a number of issues. We know that trepanation was done for epilepsy, infantile convulsions, headache and various cerebral diseases believed to be caused by confined demons to whom the hole gave a ready method of escape. While you would think this barbaric process was only done in very ancient days. We can find records and images showing the process being done in the 13th through the 21st centuries as shown below.

Additional images show how popular drilling holes in one's head really was.

Sometimes the drilling was done with special drills that had stops to ensure the brain was not destroyed, but most of the time, simply by stopping the drilling at the appropriate time. I must admit there is a variant of trepanation called craniotomy used today, where a neurosurgeon creates a burr hole to relieve the pressure of a hematoma or brain bleed. That variant is not what we are talking about here. In ancient Peru, unusual long headed skulls different from modern skulls were found. Even these had trepanned holes to fix brains as shown below.

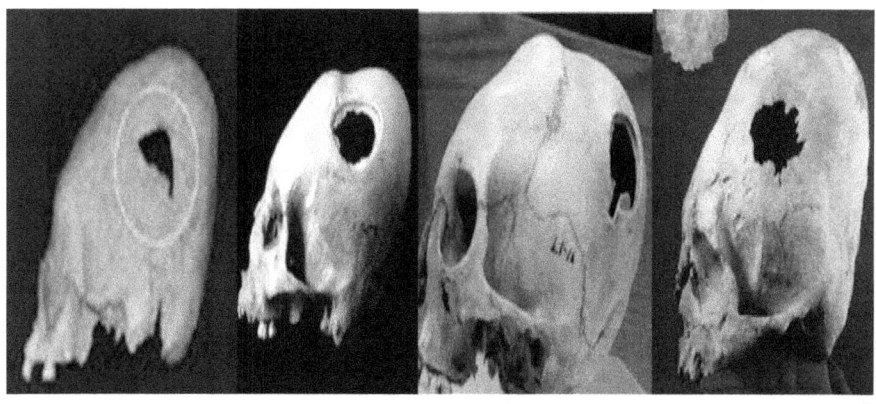

As drills were in short supply in the very ancient day, straight blade knives were used to saw through bone and matter as shown next.

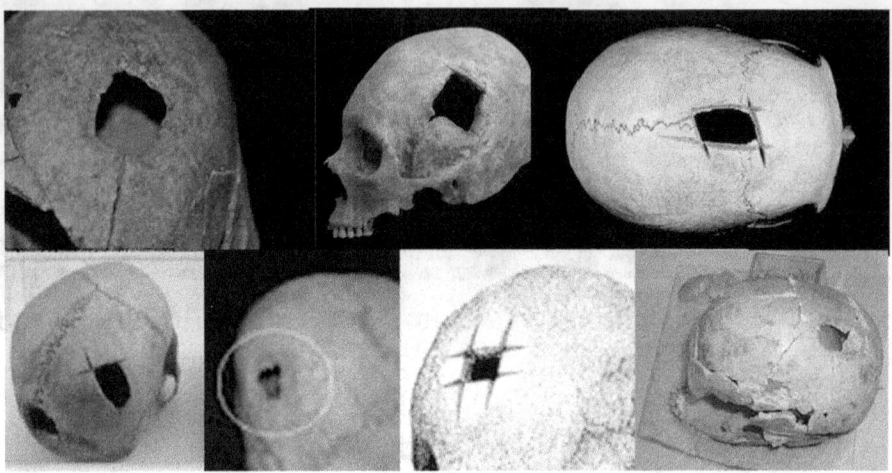

From many skulls found, it was very hard to get out of the skull what the "doctor" desired so a larger and larger hole was established. Some of these holes were enormous and evidence shows the patients did survive the process. The group following is a small sampling of the tremendous steps taken to relieve patients' concerns about their brains.

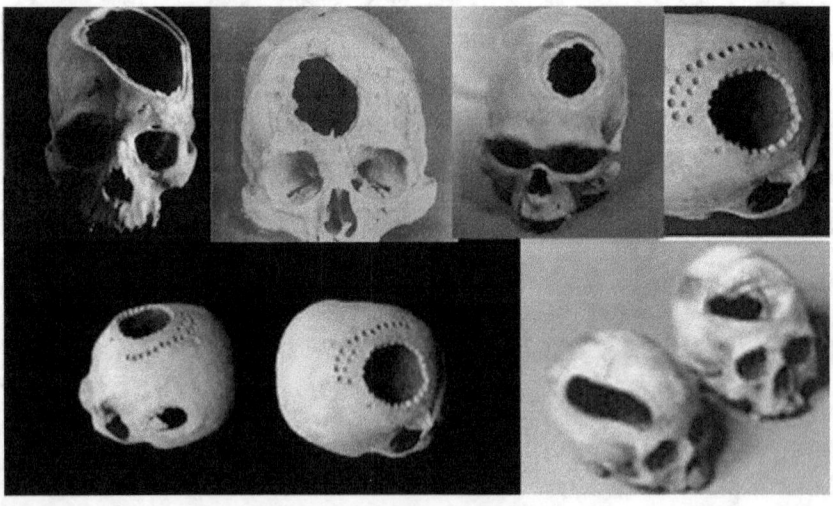

Many times, a small hole would work or many holes all over the head. Some of the skulls shown in the following group showed that the patient went back to the "doctor" many times for more holes.

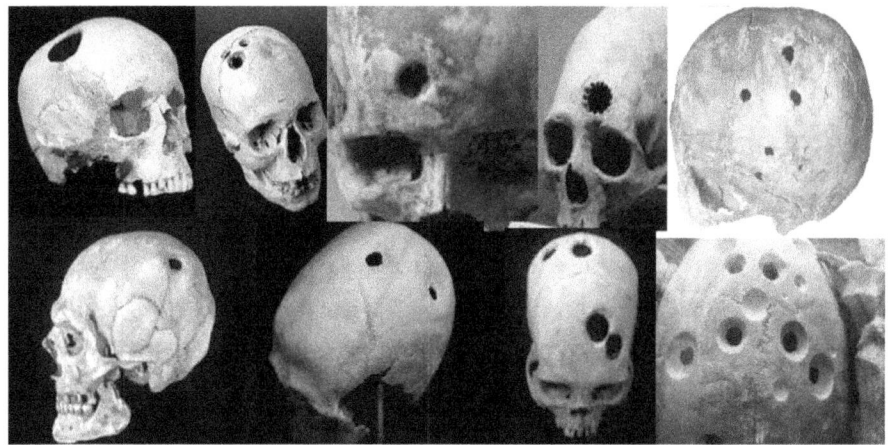

Some doctors got creative to make oval holes. Possibly this was to present a desired look after the surgery or a doctor's signature. [See the following group]

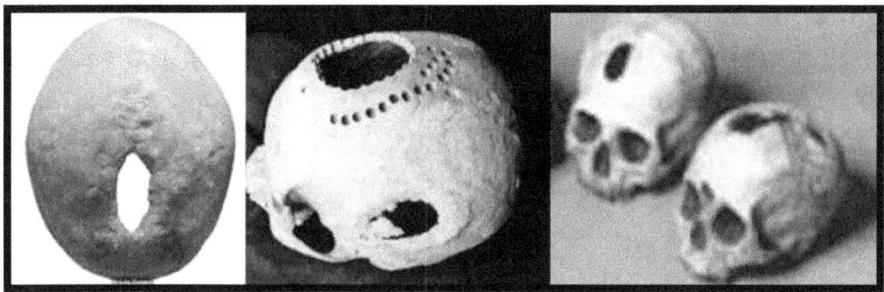

While some may have been told there would be no scaring or noticeable indication of the procedure, pictures of people who had these processes done show that the loss of skull is a noticeable characterization of those being put under the knife.

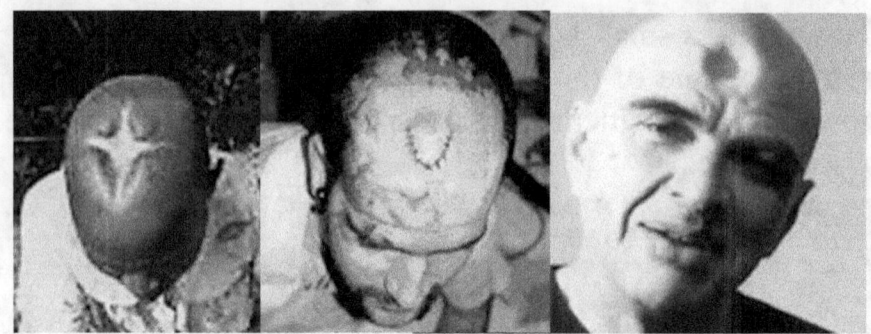

There is no way of knowing how many people lost their lives during trepanning experiments, but the practice continued because scientists told the population they would be better off with holes in their heads than demons or whatever they decided they could cure. Today, doctors certainly trepan individuals in preparation for brain surgery and for the relief of swelling due to brain trauma, while these methods are well proven by animal testing, clinical trials, and the like. A few early trepanationists simply decided to make the world better by relieving the brain from demons and the like without science. They decided by consensus and that is what we will be looking at as we go through this book. It is unbelievable how a small group came up with an idea and because of fame, backing of the leaders of a country, national pride, and other key elements of consensus science, the idea becomes more than a theory it becomes a law. Another consensus medical breakthrough was called heroine. Fake science got a real boost with Heroine.

Heroine Medicine

To save the world, doctor's leeches were replaced with Opium, Cocaine [opium], and Laudanum [opium]. Leech medication, and other old remedies were modified by consensus science into simply making people not realize they were sick to cure them. I know some doctors still use the same science and hand out narcotics to keep their patients returning, but it is not real science. Heroine started out as the cure for all coughs and ended up as the cure for everything. Below, is a tiny fraction of the varieties of opium curatives. Consensus outweighed science and observation again.

Heroine Brain Salt would cure brain troubles, sleeplessness, nervous debility, excessive headache, and sea sickness. Cocaine toothache drops would cure toothache, Blood disorders, skin eruptions, and loss of energy safely. Heroine medication called tippicanoe used the popularity of the President to cure dyspepsia, constipation, stomach disorders, trepid feelings, feeble appetite issues, female complaints, malaria, and mal-assimilation of food,

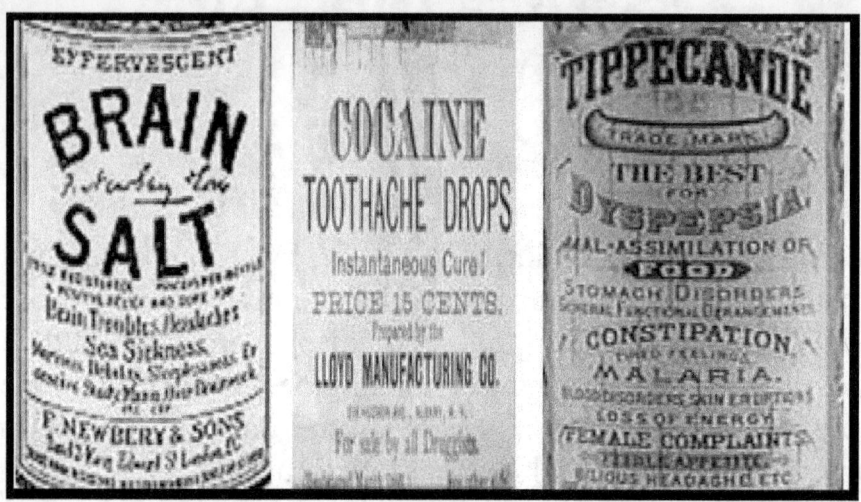

Heroine soothing syrup was marketed for children's teething. Heroine Vaporol was used as a vapor treatment for asthma and spasmodic affliction

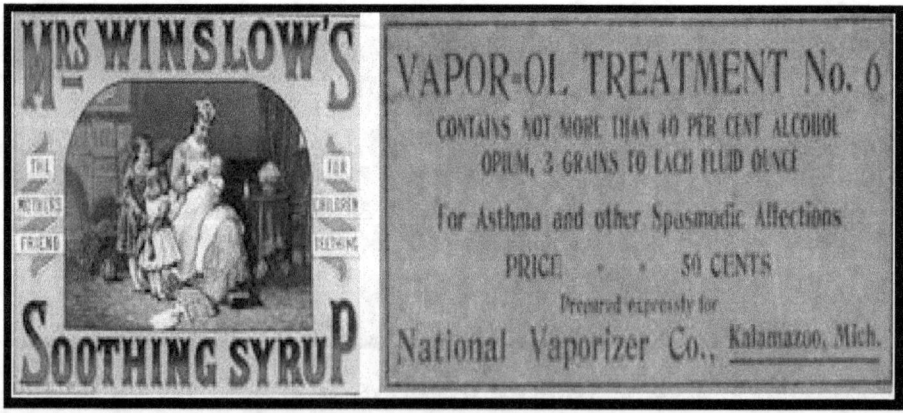

Heroine Cough syrups siad they would not affect other areas of the body and would make childrens's stomachs feel better

in a harmless way. I wonder what scientists believed harmless was.

I can remember when I was given Paragoric as a child and I had no idea they were filling me with heroine. It was called the family babysitter as it made children very docile. So that brings us to the first bayer asprin. Please notice the first Bayer asprin, middle image, was just heroin. Also shown next, heroine Daimiana medicine claimed invigoration and strengthening medication to eliminate signs of earlier in indiscretions. It invigorated both the brain and the nerves and cured impotency, and nervous debility.

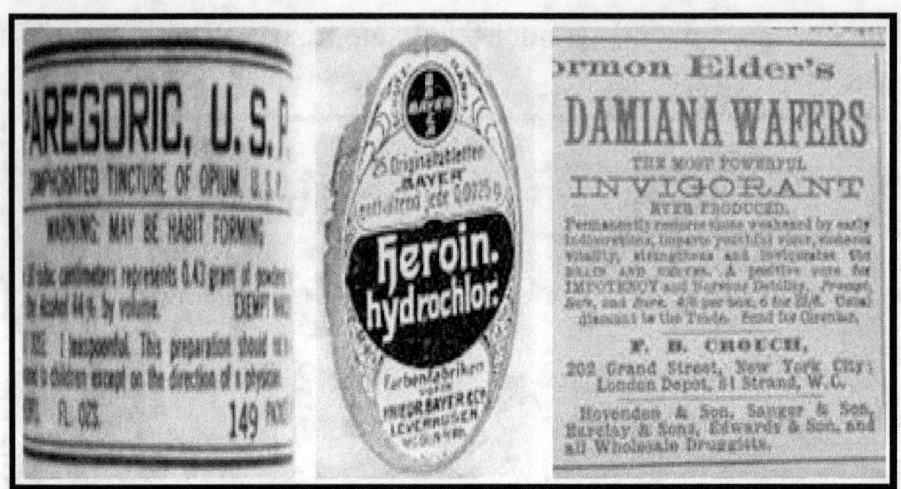

By consensus the addictive opium or heroine was widely distributed by barbers, confectioners, ironmongers, stationers, tobacconists, wine merchants, and doctors. It was easy to come by and many people took it, including numerous authors who became addicts such as Elizabeth Barrett-Browning, Lord Byron, Wilkie Collins, George Crabbe, Charles Dickens, John Keats, Percy Bysshe Shelley, and Walter Scott, Samuel Taylor Coleridge, Elizabeth Barrett Browning, and Charles Dickens. Most of them destroyed their lives just as many, many others had. Possibly the most famous of the tragedies or successes of heroine was Coleridge's partial poem *"Kubla Khan"*.

In Xanadu did Kubla Khan-A stately pleasure-dome decree: Where Alph, the sacred river, ran through caverns measureless to man- Down to a sunless sea. So twice five miles of fertile ground with walls and towers were girdled round: And there were gardens bright with sinuous rills, where blossomed many an incense-bearing tree;And here were forests ancient as the hills, Enfolding sunny spots of greenery. -------

His written "partial poem" obtained from an opium stupor is shown next. He would spend the remaining 30 or so years as an addict and even went to live with his drug supplying doctor for the last 18 years. To the right a special form of opium with Lead used to get an extra kick. If the opium didn't get you the lead poisoning would. It was very popular in Coleridge's England.

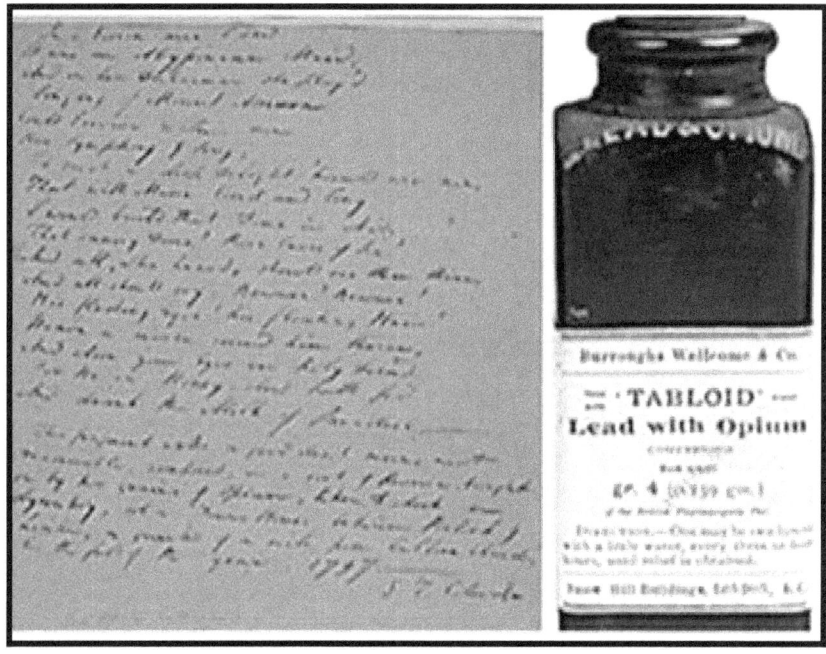

There was nothing heroine could not "cure". Soon just about everyone had heroine on hand at home to cure just about everything. Patients loved how well the medicine worked and came back over and over again to be cured. Today we have names like Oxycodone, Hydrocodone, and Fentanyl to hide the dangers of opioids and doctors are still treating patients by addiction rather than curation. The entire world looks better while medicated so it must be curing the entire world. As heroine lost favor, consensus scientist grabbed on to a new way to control the masses and give themselves power. This was done with Radium.

Miracle of Radium

I know you are thinking the Heroine consensus was the worst, but it wasn't even close to this next one. Consensus doctors and practitioners decided the entire world should grab hold of the miracle known as Radium. Home and doctor use of Radium curatives and treatments abound through the first half of the 20th century and beyond as consensus scientists developed cures for anything heroine had problems with. Radium was sprayed, smeared, injected, pushed down every orifice you could imagine as the consensus scientists told the world that they would essentially die without this miracle curative. I know you hear about things that are going to destroy the world by these same types of sinister people controlling entire populations by fear as they parade some TRUTH that their group made up to get them rich, famous, or even to promote some fake scientific agenda. Radium scientists, engineers and manufacturers had their day during the first half of the 20th century. Degnen's Radio-Active Solar Pad purchased by tens of thousands would eliminate sluggishness, high blood pressure, diseases of the stomach, heart, lungs, liver, kidney, constipation, gout, sciatica, and other similar ailments. For Men, there was the "Testone Radium Energizer and Suspensory". [2nd below] "The Rectorotor" came along to add radium through the butt to cure constipation, prostatitis, piles, and similar discomforts. It was decided that Radium could cure just about anything.

Radium suppositories would assure a better sex life, as shown next.

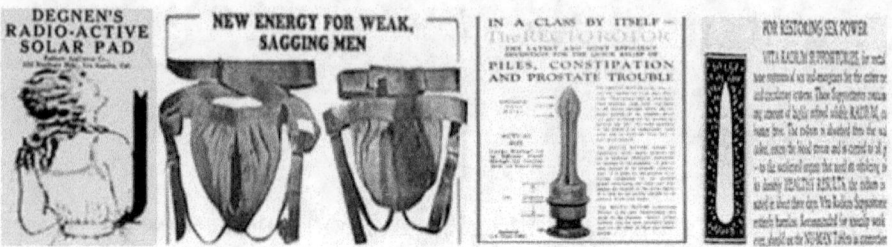

A tube was inserted to deliver radium cleansing for Bladder and Urethra issues as shown next left. Other radium treatments had the hands and feet placed in a radioactive bath. Sometimes, just breathing radium in the air could fix your woes as shown next right.

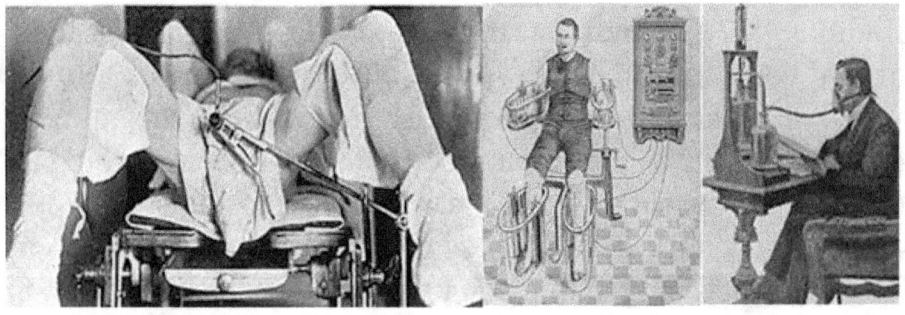

For skin Cancer they came up with radium location modules for the skull or as a partial mask on the jaw [see last 2 images].

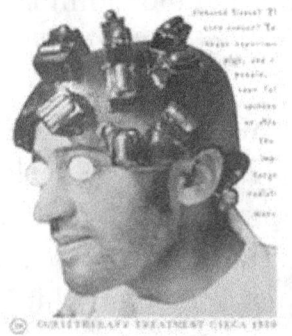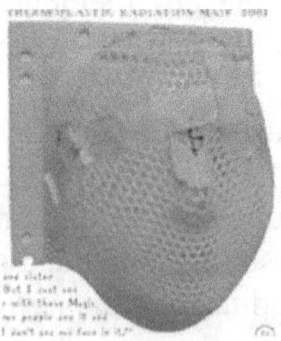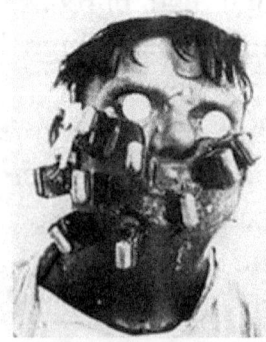

"Radithor" would cure Arthritis, Rheumatism, Neuritis, Gout and other joint issues. Radium slave could be spread to eliminate cancerous growths. Tho-Radia made a whole product line of perfumes, creams, facial powders, lipsticks and other beauty products that contained thorium chloride and radium while an interesting curative was the Radium chocolates by Burkbraun. Radio-Sulpho would dissolve all poisons from the body as a germicide, antiseptic, pore cleaner, and eliminated urinemia, Rheumatism, Swellings, Blood poisons, and, of course, Cancers.

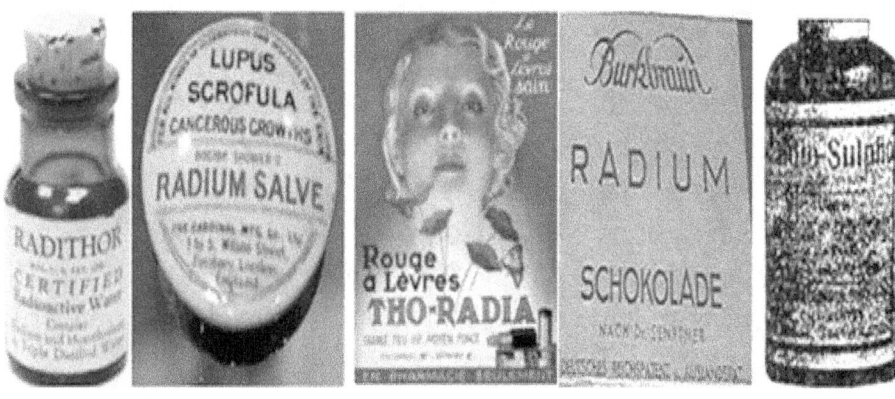

Radium Condoms by NUTEX assured a high percentage of accuracy. This medical device was patented in 1912 by R. W. Thomas and manufactured by the Radium Ore Revigator Co., which sold thousands of them in the 1920s and 1930s. It was basically a ceramic water crock lined with radioactive materials (uranium, vanadium, radon), lead and arsenic. Radium Hand Cleaner – "*It Takes Off Everything but The Skin*". La Parle Obesity Soap even eliminated fat with Radium. Drinking water was replaced with Radio-active radium infused drinking water to cure "*Tabes Dorsalis*", "*Catarrh of the Antrum and Sinus*", Diabetes, Glycosuria, Nephites, and other ailments in the hospital and home using the third device. If you didn't have a home fountain, you could use Radione tablets to add to water for energy. Radi-endocrinator papers were spread over the inflicted area to fix the endocrine glands.

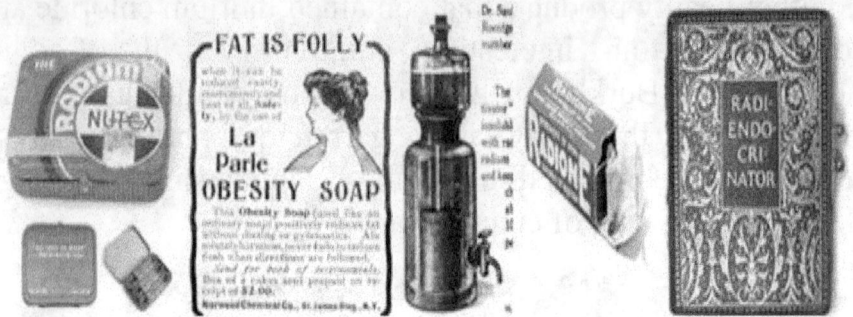

Glen Springs Health Resort and Hotel opened up as a health resort with its highly radioactive Radium hot springs. Just relax and cure everything. At home one would simply use Radio-active radium "Xray Soap" and radioactive Laradium face-cream made the complexion a thing of beauty in just 20 minutes. Radium Hand cleaner was advertised as being able to remove everything but skin.

Radium tablets under the name brand Arium simply claimed to make you a super-man. Radioactive wool by Laine Oradium for self-heating clothing and Radium heating pads kept everyone warm and safe. Then there was "Caradium hair conditioner" that would eliminate grey.

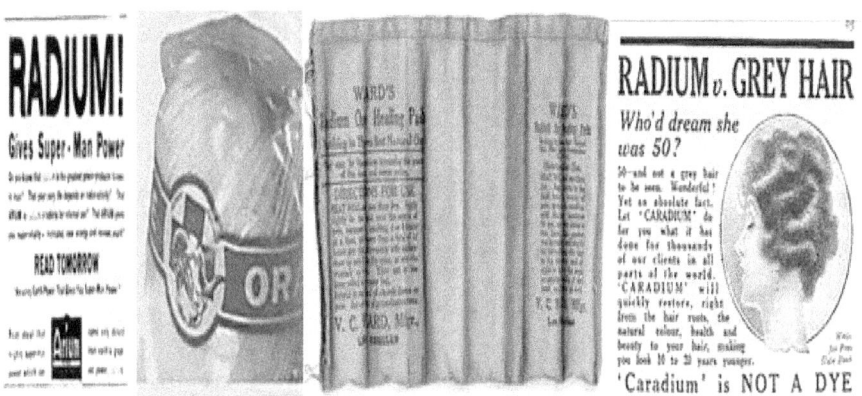

For Eye Problems the consensus scientists came up with "Radio-bleu while Radio-X tablets could clean the bile ducts. Even teeth got brighter using Doramad radium toothpaste. For Smokers we find radium infused "A-Batschari" Cigarettes to be lit by Radium "silent matches".

Multiuse "Radium Spray" could be used to polish your furniture and if a bug fly by you could kill it with the same spray. Almost the same Radium formula was marketed as leather dye and as metal polish. Radium "Gloss Starch" could even clean your clothes.

For Lunbago, Sprains, Bruises, Chest colds, swollen joints, and coughing, just take a little "Radium Radia". There was even reports of Radium bringing back sight and doctors would routinely use a radium compress or as an intravenous injection shown third and 4th.

The X-Radium foot warmer came along with a huge number of commercial aids including glow-in-the-dark radium painted chandeliers, glowing jewelry, and glow in the dark wrist watches. I had one of those things when I was younger myself. Even glow in the dark Roulette became the rage.

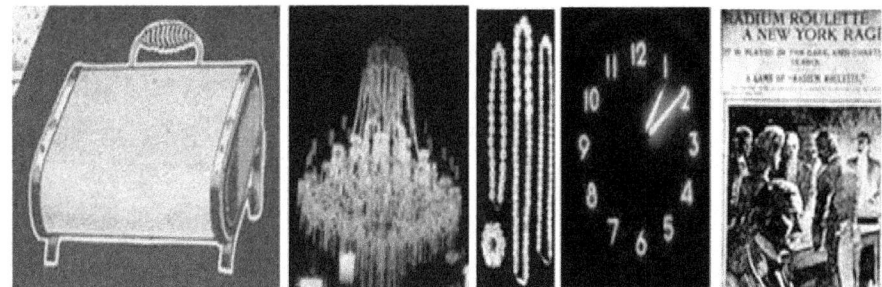

Soon radium death tolls began to climb. Those not dying immediately were sometimes disfigured horribly. I even got rid of my radium watch. You would think people have more sense than to drink something that was being sold as a bug killer, but consensus scientists can sway just about anyone and a new consensus science was just around the corner.

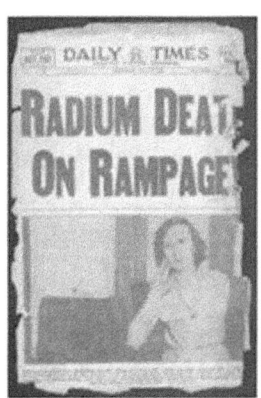  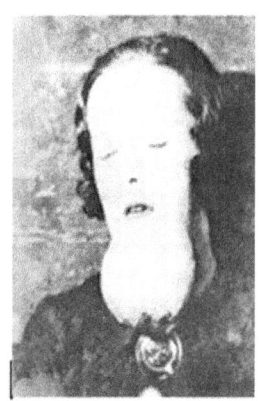

Science by consensus had claimed many more victims and all because we are too trusting when someone puts scientist before their name. Before the Radium scientific medications, we had bile science. Surely scientists would not simply make things up to make them seem important.

Fake Biology

For this section we will go back to the time of the Greeks, to look at a consensus "known characteristic of our biology" that would govern medicine for over a thousand years. This was initiated by the father of medicine, Hippocrates.

Greek physician Hippocrates (460 BC to 370 BC) is often credited with developing the theory of the four humors. In this well tested theory, it was established that the body was run by these humors; blood, yellow bile, black bile, and phlegm, and their influence on the body and its emotions was substantial. If you had too much phlegm, for instance, your body would have some issue or your character would be manifested in a negative way. Too much blood could be corrected by bleeding or later by adding leeches to a wound to correct the destabilizing humor. As shown below humors worked in consensus with the 4 elements of the Earth, Fire, Air, and Water [whatever that meant.].

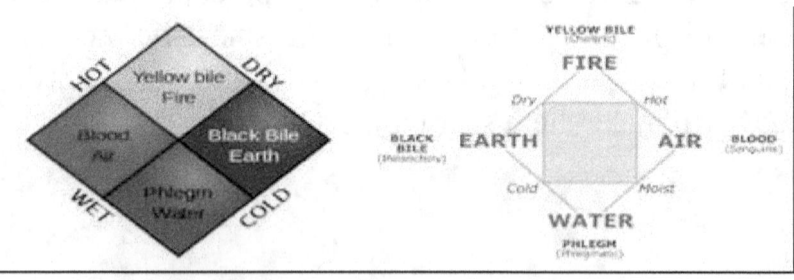

As you would expect from it well tested correctness, this science remained by consensus right up to modern times. In this science, it was believed that the right balance of these four humors made a person healthy but an excess or decrease in any one of these would cause illness. Because of this belief, treatments of sickness would include bloodletting, purges, and emetics. Almost all medieval medical treatments were designed to cure by restoring the natural balance of the four humors. Doctors knew that a person's health and personality were dictated by the 4 humors as shown below.

Sanguine personality- *blood controlled-* happy, generous and optimistic, but irresponsible personality.

Choleric personality- *yellow bile controlled-* violent and short tempered, but ambitious.

Phlegmatic personality- *Phlegm controlled-* Sluggish, pallid, and cowardly

Melancholic personality- *Black bile controlled-* Introspective and sentimental

Other definitions are captured in the following graphic.

Humor	Season	Element	Organ	Qualities	Ancient name	Modern	Ancient characteristics
Blood	spring	air	liver	warm & moist	sanguine	artisan	courageous, hopeful, amorous
Yellow bile	summer	fire	spleen	warm & dry	choleric	idealist	easily angered, bad tempered
Black bile	autumn	earth	gall bladder	cold & dry	melancholic	guardian	despondent, sleepless, irritable
Phlegm	winter	water	brain/lungs	cold & moist	phlegmatic	rational	calm, unemotional

Occasionally a mixture of herbs would be used to restore the balance. The humors were also applied to foods – for example wine was choleric (yellow bile). This classification still exists today to some extent, as we refer to some foods as "hot" and others as "dry". To treat melancholy and mental illness, treatments included the manufacture of blisters to regulate humors and forcing vomiting. Vomit machines were used successfully [just ask the consensus scientists].

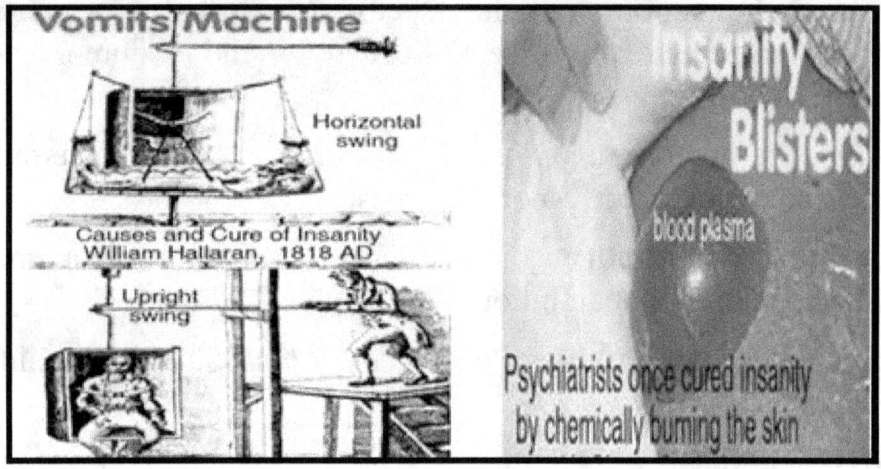

The concept of humors was not replaced until 1858 when Rudolf Virchow published theories of cellular pathology. To make the world a better place, we also have consensus scientist that used Anthropology to save the world from de-evolution. If you think I am not a fan of consensus science over real science, you would be absolutely right. The following sections may help you feel the same way.

Fake Aryanism

I think a great example of consensus science is something called Aryanism. This takes us back to Nazi Germany. While there was some negative distinction about being a black man and the thought that a negroid person was less evolved than normal people; we will look at that sort of consensus Evolutionism shortly as this **"science"** was substantially different. This was the desperate attempt to push out or eliminate the round headed Jewish populations to save the country of Germany and eventually save the entire world. This public service was initially established by head shape, but later additional criteria allowed for expanded exterminations.

Georges Vacher de Lapouge-[1890] - He came up with a somewhat new concept in his book *"The Aryan and his social role"* claiming the superiority of the Aryan race and divided humanity into various, hierarchized, different "races", spanning from the *"Aryan white race, dolichocephalic"*, to the *"brachycephalic mediocre and inert race"*, best represented by the "Jew." Dolichocephalic simply means having a long thin head while brachycephalic is a round headed, low class person. This included the "*Homo Europaeus* (non-Aryan Teutonic and Protestant), the "*Homo Alpinus*" (French Auvergnat and Turkish), and finally the "*Homo Mediterraneus*" (Spanish and Italians.)." *Homo Africanus"* (all black people) were not even considered people. Vacher de Lapouge became one of the leading inspirations of Nazi anti-Semitism and Nazi ideology. Some even suggest he sparked the fire for the extermination of the Jewish Poles. Here is an excerpt of his writing.

In the next century, people will be slaughtered by the millions for the sake of one or two degrees on the cephalic index. That will be the sign, replacing the biblical Sabbath and the linguistic affinities that are now the markers of nationality, Only it will not have anything to do, as it does today, with questions of moving frontiers a few kilometers; the superior races will substitute themselves by force for the human groups retarded in evolution, and the last sentimentalists will witness the copious exterminations of entire peoples.

William Z. Ripley [1900] - He came up with almost identical characterizations with the in *The Races of Europe.*

Anders Retzius [1920]- He used the thin headed Swedes and the Aryan master race and the bad round headed Slavic people as the low-level people, but the story was the same.

The German people had no work, no money and were starving. A wheelbarrow full of 100 billion-mark banknotes could not buy a loaf of bread at the time, and many Germans were living in shacks after countless homes and farms had been seized by the Jewish control Rothschild/Rockefeller world banks. After Hitler was elected in in 1933, he refused to play ball with the Rockefeller-Rothschild rules and completely thwarting the international banking cartels, many Jewish bankers were killed and, as a result, the Nazi government issued its own currency known as Reich Marchs, which were debt free and uncontrollable by international financial interests. Inflation stopped immediately. Everyone cheered and life got back on track. Hitler noticed that all the people causing the inflation and destruction of his country were Jewish so he talked to his scientists and the ones who wanted to stay alive and prosper told him that Jews were a less evolved group than the Aryan race. Praise and adoration came so another claimed the Aryan features of straight nose, high flat forehead etc. could be used to determine which people should or should not procreate to save Germany and the rest of the world. The consensus science community strongly advised Aryan marriage to Aryan Germans to support building a stronger country and money was given for each Aryan child. If a scientist disagreed with Aryanism, he was fired or worse just like those who go against what has been called "human caused global warming in America". Soon almost all scientists accepted and praised the Aryanism assertion just like all consensus scientists do as the want to get government grants, write award winning technical papers, get fame, and keep their jobs. Thousands acclaimed the Aryanism science to be great and those not agreeing were in some way against the country. Papers were written, showing how the Aryan race would be able to save the world from destruction of de-evolution. Time went on and Hitler didn't like the Polish people for one reason or another even though they looked Aryan. Scientists set out to establish reasons why the Slavic people should be kept from procreating even with a flat head as the government decided to destroy all 3.5 million of them so the science community pushed the Aryanism to include all Jews as there was a baseness that could not be allowed and the Polish Jews would de-evolve Germany. They had become part of the great awakening to save the earth. Some Aryan looking Jewish women were given a reprieve from gas chambers and the like by allowing them to marry non-Jewish Aryan men. Because they evidently believed in Aryanism, they went along and married the men.

Nazi social policies and German consensus science placed the improvement of the Aryan race through Aryanism at the center of Nazis ideology and the most effective method to save the world. Those humans were targeted who were identified as "*life unworthy of life*", including but not limited to Jewish people of Poland and everywhere else, slowly began to include criminals, degenerates, dissidents, the feeble-minded, homosexuals, idle, insane, and the weak. For the sake of humanity, they were scheduled for elimination from the chain of heredity. Despite their still looking Aryan, Nazi scientists agreed that extermination of Slavic heritage (Poles, Russians, Ukrainians, etc.) could only be used as slaves or set up for extermination.

Don't get me wrong, there was no "true" scientific reasoning concerning Aryanism, but consensus science won out for the sake of saving the world just like it does every time it grabs hold of a society as we find in the notion that human caused global warming is by something as crazy as CO_2 when H_2O causes the most devastating changes in our atmosphere and hemostasis. In this case racial bias and some innate hatred for Jewish people and national pride drove the Aryanism science. For the human caused global warm-ists the drive started out as concern for a warming trend that warped into simple greed and power-lust. Before we get into that subject, let's look at other similar consensus science that pushes ideals without science and appears scientific simply because "scientist" help push the idea just like in Nazi Germany. The next consensus is with something called Polygenism which was actually a desire to eliminate the de-evolution caused by black people. Should we question Scientists?

Scientific Polygenism

Similar to Aryanism was a consensus science called Polygenism. This science was going to save the civilized world from barbarians by establishing classification of people around the world. The "quasi-scientists" simply started out with a "known concept" that black people were somehow less human and all of a sudden, a new consensus science was born. Some say Voltaire was the inventor and reason behind the "science".

Voltaire 1734- He wrote a book *"Traité de métaphysique"*. In it we find the following "*Whites, Negroes, and the yellow races are not descended from the same man*". He believed each race had separate origins because they were so racially diverse. To show how diverse he wrote, *"It is a serious question among them whether the Africans are descended from monkeys or whether the monkeys come from them. Our wise men have said that man was created in the image of God. Now here is a lovely image of the Divine Maker: a flat and black nose with little or hardly any intelligence. A time will doubtless come when these animals will know how to cultivate the land well, beautify their houses and gardens, and know the paths of the stars: one needs time for everything.* When comparing Caucasians to Negros, Voltaire claimed they are different species: *The Negro race is a species of men different from ours as the breed of spaniels is from that of greyhounds. The mucous membrane, or network, which nature has spread between the muscles and the skin, is white in us and black or copper-colored in them.* Voltaire is shown next left.

1787-Samuel Stanhope Smith – He was an American Presbyterian Minister and author of *Essay on the Causes of Variety of Complexion and Figure in the Human Species* in 1787. Smith projected that Negro pigmentation was *"nothing more than a huge freckle that covered the whole body as a result of an oversupply of bile, which was caused by tropical climates"*. Later we will talk about this bile stuff as many believed during this time another Consensus presented around 380BC that people were made up of equal parts of blood, yellow bile, black bile, and phlegm so long as all these things stayed regulated, a person would remain healthy and intelligent. *"By Negros having too much bile it messed up their intelligence."* Certainly, a Christian minister would not use consensus science to destroy what people thought about black people, but there he was as shown next middle. I know you are thinking the science community could not go along with this untested, undeserving "theory", but you would be wrong and just as sad as me as George Culver comes along, see next right.

1800-Georges Cuvier- This guy had all the credentials the French naturalist and zoologist. Rather than using some testing and verification he used his understanding of nature and zoological characteristics and the influence of the established scientific Polygenism of Smith. Cuvier "reasoned" there were three distinct races: *the Caucasian (white), Mongolian (yellow) and the Ethiopian (black)*. Each had its place in the world and should be rated by the beauty or ugliness of the skull and quality of their civilizations. Cuvier established the scientific polygenetic understanding of humans.

<u>Caucasians</u>: *"The white race, with oval face, straight hair and nose, to which the civilized people of Europe belong and which appear to us the most beautiful of all, is also superior to others by its genius, courage and activity."*

<u>Negros</u>: *"The Negro race is marked by black complexion, crisped or woolly hair, compressed cranium and a flat nose. The projection of the lower parts of the face, and the thick lips, evidently approximate it to the monkey tribe: the hordes of which it consists have always remained in the most complete state of barbarism."*

1850- Ernst Haeckel – This consensus scientist, in his doctrine of evolutionary Polygenism, wrote that *"Negroes have stronger and more freely <u>movable toes</u> than any other race which is evidence that Negroes are related to apes because when apes stop climbing in trees they hold on to the trees with their toes"*. He established black people as *"four-handed apes and savages"*, classifying Whites as *"the most civilized."*

Unbelievably, he never had a white man and a black man tested for toe motion; it was simply stated by consensus and no one tested this bile stuff or had any inclination of what it might be.

Consensus Monogenism

The other side of the Polygenism was called monogenism or the Out-of-Africa truth. For this one the proclamation came that Adam was the first man and the first man came from Africa so Adam was a black man and we split away from his initial coloration. Initially DNA testing or Haplotyping seemed to go along with this absurdity and mapping showed how everyone came for homo-erectus. Later, white people came from India around the time of the Pleistocene Extinction as an offshoot of what now was the Neanderthal. It should be noted that Neanderthal have never been found in India and flooded the European countryside during the Pleistocene.

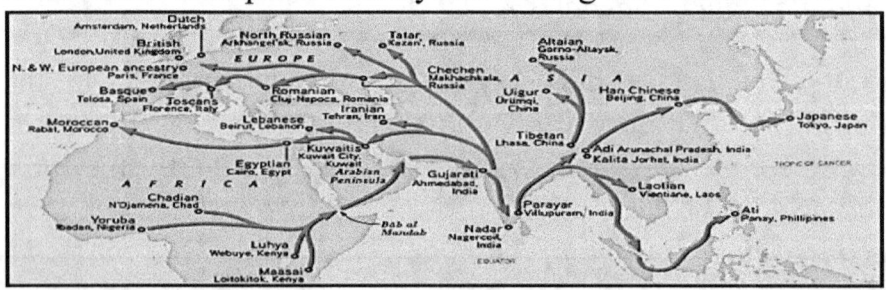

Soon it was found that while homo-erectus based people did indeed come out of Africa, a brand-new type of man spontaneously was generated in the Middle East. The African based people began as black people and were of a much older race than the redheaded Middle Eastern based group. By the time this new human came along, the DNA suggests that the African and Neanderthal humans had spread to Europe and

Asia without there being signs in the new DNA humans of the Middle East.

Rather than the "Out of Africa map, we might see the following with the new data.

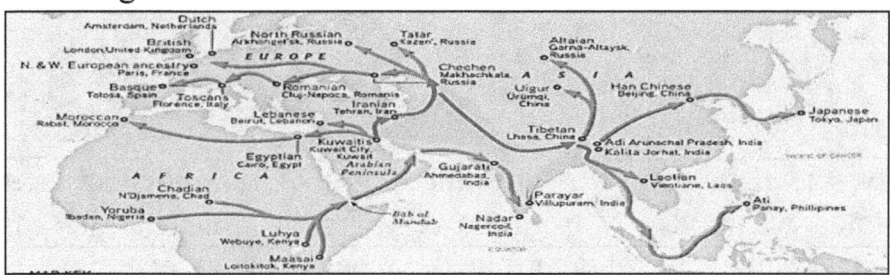

Even with the new information, text books continued to show that all descended form Homo-Erectus and came out of Africa. In Egypt, monogenism Scientists began changing the names on statues to show Egyptians were black men. To sort of give you a feeling for how hard people try to place the Egyptians with Nubia, the following statues are both reported to be the Theban king during the 11th dynasty, Amenemhat III. The first is easily determined to be Caucasian while the second one is certainly Nubian. By the way, these cannot be the same king. Nubian Kings took control of Egypt briefly after the time of King Solomon who provided protection for gold helped them seize and hold onto control until the Assyrians finally took the country a little over a thousand years ago. One can assume the king to the right is one of the much later Nubian kings of the 22nd through 25th dynasties. I know you are thinking people in Egypt hate noses and smash them, but ignore that.

By scientists trying to show the Out-of-Africa phenomenon, almost all the Rulers from the very first "Narmer" until King Tutankhamen were determined to be black at one time or another. Below, is an incomplete list of those determined to be black. A number look black alright, but most don't have strong "black" features at all and other examples of these same rulers have completely different features as if someone was trying to drive home the possibility that ancient Egyptians were all black.

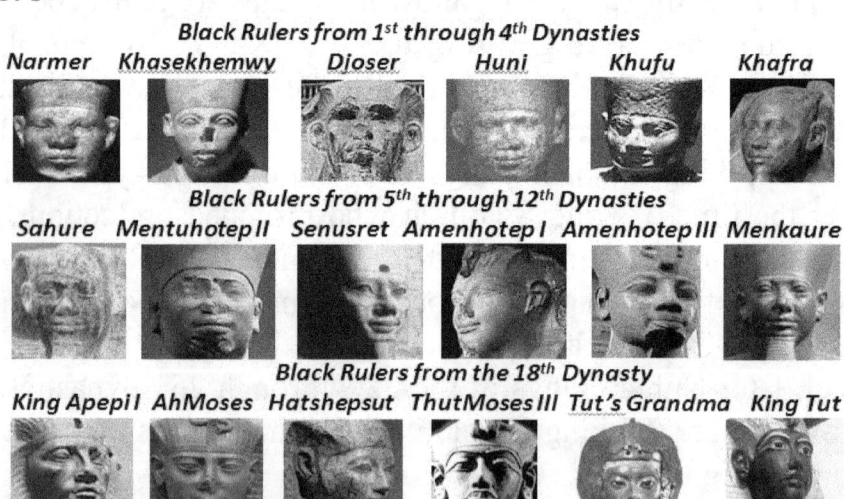

Below are the white images of Mentuhotep II, Amenemhet III, AhMoses, ThutMoses III, and King Tut---before sinister consensus scientists turned them into black people to protect the world.

Hopefully, you are beginning to realize, it is not the theory that causes such great harm by the consensus scientists, it is the lengths they will go to prove their statements without experimentation, theory testing, or reason and how entire societies can so easily be convinced these guys are telling them truth simply because they are scientists.

Fake Phrenology

This FAKE science of Phrenology with a bit of Craniometry mixed in was made famous by Cesare Lombroso, the founder of anthropological criminology around 1880. He claimed to be able to scientifically identify links between the nature of a crime and the personality or physical appearance of the offender. This is what some have called "born criminal" unlike some of these guys, Lombroso did do some level of experimentation, but he was so bought into the result he had already introduced, the experimenting did little to modify the theories. He concluded that skull and facial features were clues to genetic criminality, and that these features could be measured with craniometers, goniometers, and calipers, as shown below, with the results developed into quantitative research.

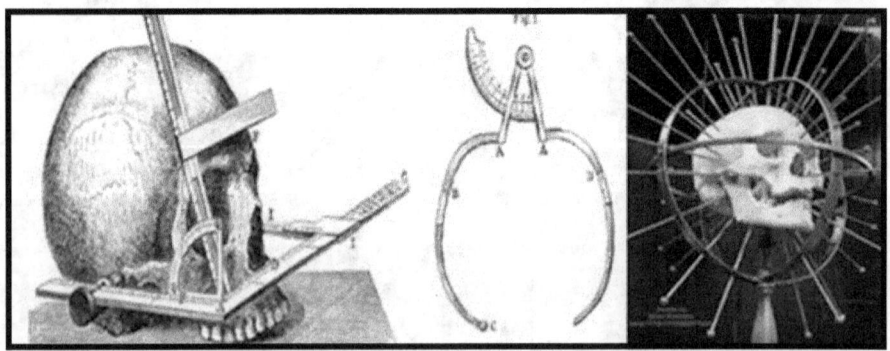

A few of the 14 identified traits of a criminal included large jaws, forward projection of jaw, low sloping forehead; high cheekbones, flattened or upturned nose; handle-shaped ears; hawk-like noses or fleshy lips; hard shifty eyes; scanty beard or baldness; insensitivity to pain; long arms, and so on. Here is a news article of how this nonsensical science would help mankind classify criminals.

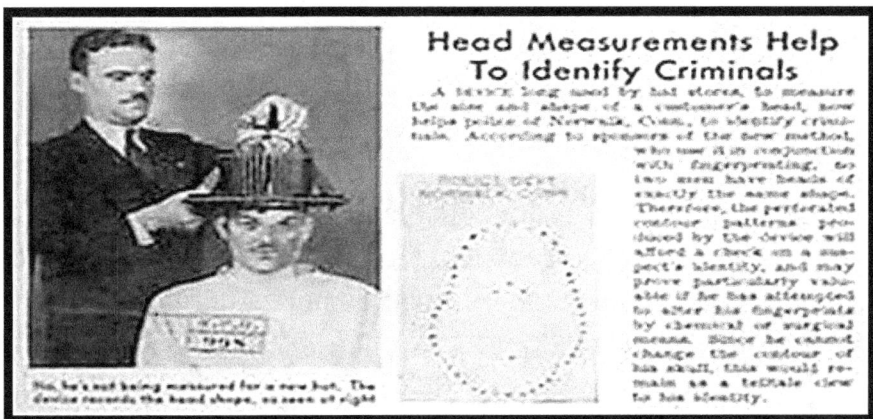

Massive devices full of test points assured the testing would be thorough and unbiased. This even extended to electronic measurements as shown below right.

Facial Angle Testing-Pieter Camper [1750] had distinguished both as an artist and as an anatomist, published some lectures containing an account of his craniometrical methods. These laid the foundation of all subsequent work. He was the inventor of the "facial angle", a measure meant to determine intelligence among various species. According to this technique, a "facial angle" was formed by drawing two lines: one horizontally from the nostril to the ear; and the other perpendicularly from the advancing part of the upper jawbone to the most prominent part of the forehead.

While this was also used in the Aryanism attempt to cleanse the world his ideas continued throughout the 19th and into the 20th centuries. Camper claimed that antique statues presented an angle of 90°, Europeans of 80°, Black people of 70° and the orangutan of 58°, thus displaying a hierarchic view of mankind, based on a decadent conception of history as we became more and more de-evolved. The testing became more and more involved as shown.

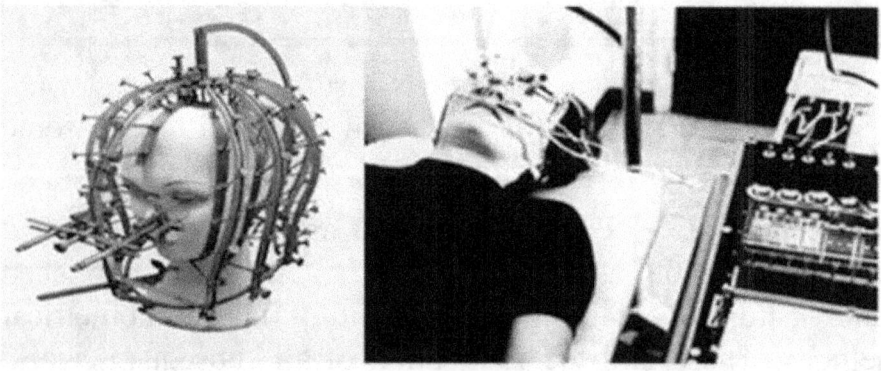

This consensus scientific research was continued by many others using many complex devices to assure the testing was accurate and "scientific"; then the bumps came in.

Craniometry-It wasn't long before everyone thought people could read an individual's personality, their strengths and weaknesses, hopes, intelligence, criminality, and desires, by examining the pattern of bumps on their skull. Craniometry or reading bumps on your head help with a strange vison of the brain. They say it is lumpy and where the brain had lumps, the skull would follow. Therefore, by measuring those bumps, one can infer which parts of the brain are enlarged and therefore which characteristics are dominant.

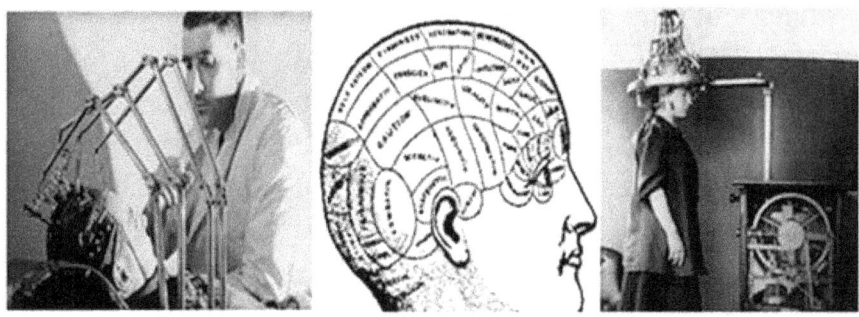

By the middle of the 19th century, phrenology parlors were widespread. Automated phrenology machines came later. The automated machines were composed of numerous spring-loaded probes. The device was placed over the head while the probes would extend to gently touch the scalp, thereby providing a measurement of the topography of the skull. The machine would then calculate the characteristics of the subject based upon this topography and produce an automated reading. One could even test himself using devices shown below.

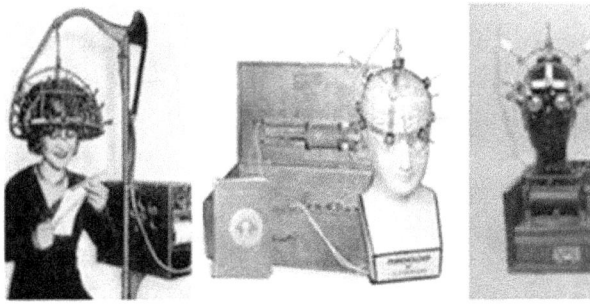

This stupidity continued. In the late 19th and early 20th centuries, the concepts of phrenology became associated with consensus scientific ideas in the areas of criminology and evolution that were popular at the time. Craniology and Anthropometry were attempts at identifying evolutionary advancement and criminal tendency according to physical measurements of the skull and face. As with most of these consensus sciences, these measurements were used to verify pre-existing social prejudices. To give the testing an air of reasonableness, automatic and robotic tasting machines shot up around the world as we tried to protect ourselves from ourselves. Here are more examples of robotic testing systems that assured the results could be help the world understanding of its inhabitants.

  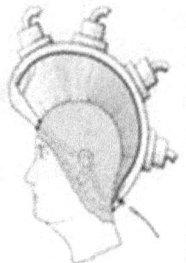 

All the Phrenologist scientists were trying to protect the environment of the world by weeding out humans who should not procreate. We can imagine many found their nose angle was too long and would not marry or a bump on their head showed they would commit crimes and made them not try for children so the Earth was saved. Speaking of bumps; we need to talk about moles.

Fake Moleoscopy

Unbelievably, consensus science claimed a victory in medicine again with something called moleoscopy. The study of moles would determine the nature of a person and active practiced of this began in the 18th and continued as late as the 19th century. Doctors would simply check the skin for any moles and help the patient. These classifications could be used to help or hurt getting a job, winning a case in court, or finding the right person to marry. Today many have broken free from consensus on this one, but at one time here is the extensive characterization controlled by moles.

Rib Mole—A mole on left rib away from the nipple shows a lazy person

Ankle Mole — On a man this shows a fearful nature. On a woman, a sense of humor, courageous; is willing to share love and worldly possessions with others.

Armpit Mole— Under the left arm indicates the early years are a struggle but ample remuneration, even riches will make the later years very happy. Under the right arm this means constant vigilance for welfare and security must be kept uppermost.

Back Moles- If you have a back mole you must be sure you have all the facts before you enter into negotiation of any enterprise.

Belly Mole– Belly mole people have a tendency to self-indulgence. Avoid overeating and excessive drinking. Keep a rigid check on your economy. Choose a marital partner who has an even, calm temperament and the gift of understanding.

Bosom Mole—With a quarrelsome nature given to temper; this mole indicates a lazy, sometimes unsteady disposition. Lack of ambition may result in a colorless career. On a man, this area is the chest.

Breast Mole – A mole on the right breast, indolence and intemperance may destroy happiness for self and family. Need to exert will power and self-discipline so that you can enjoy the love and comfort the children you might have. A mole on the left breast made a person active, energetic, and able to concentrate upon the acquisition of wealth and property.

Nipple Mole-- On a man it indicates he is fickle and desirous of many amours. On a woman it shows she is always striving for social status.

Nose Mole — This shows someone who will be a sincere friend. Will achieve success, make an excellent marriage. A person dedicated to amassing great wealth even though the struggle often seems impossible. Mole on bridge of nose-A lustful person

Buttocks Moles—This shows a person is not very ambitious and inclined to accept any mode of living, even poverty.

 Ear Mole-- Rare, but whoever possesses it may find riches far beyond expectations.

Elbow Mole—This mole shows a tremendous desire to travel and always uncertain. Usually talent connected with one or more of the arts. He is capable of earning a fortune in money but rarely having the urge to work for it.

Eye Mole — Poverty overshadows unusual talent from this indicator. If the mole is located on the outside corner of the eye it means an honest, forthright person, one who is reliable but needs love and admiration to offset the struggle for existence.

Eyebrow Mole — Over the right eyebrow, a mole signifies perseverance and a very active life and successful in everything — business, home, and family. Over the left eyebrow or temple, the reverse is threatened. Disappointments will be due to selfishness and indolence. Only with a maximum of effort can poverty be avoided.

Finger Mole — On any finger, this shows dishonesty, inclined to exaggerate due to inability to face the hardships that must be confronted.

Foot Mole—This shows someone who is inclined to brooding akin to melancholia. Prefers a sedentary life but really needs a balanced amount of activity to remain healthy.

Forehead Mole-- The middle of the forehead, a mole predicts honors, wealth, love, and a happy, distinguished family. Right and left forehead mole interpretations are identical with "eyebrow" classification.

Groin Mole – Despite prosperity, a mole on the right groin shows a propensity for ill health. On the left side, this type of mole shows frailty without much prosperity.

Hand Mole — This shows an abundance of almost everything, health, wealth, and happiness. Usually this person is very talented. Some tattoo on hand moles to see if their lives will change.

Heel Moles— People who are very active mentally and physically have heel moles. This shows the ability to accumulate a fortune if so inclined, but makes enemies who continuously plague and cause petty annoyances.

Hip Mole — A mole on any part of the hips except the buttocks, contentment, fortitude, and ingenuity are the salient attributes that balance an otherwise over-amorous nature.

Instep Mole – Instep moles show quarrelsome nature. They are often sullen and generally have a keen interest in athletics.

Knee Mole — On the right knee we find a friendly, amiable disposition; a great lover, desirous of family and home life. On the left knee we find extravagant and inconsistent nature, but an excellent business acumen.

Leg Mole— This shows many difficulties during early years but capable of surmounting them by sheer forcefulness. Resources must not be dissipated. Avert any tendency toward indolence.

Cheek Moles--On either cheek this mole shows a serious, studious, almost solemn person. They have a middle-of-the-road point of view on most theories pertaining to living, religion, and politics and consider wealth as not necessary for happiness.

Lip Mole— This person has a benevolent nature, always striving for better conditions. [See below left]

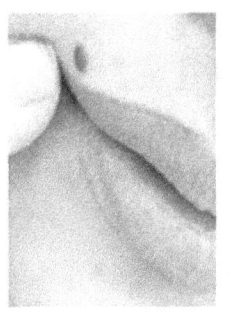  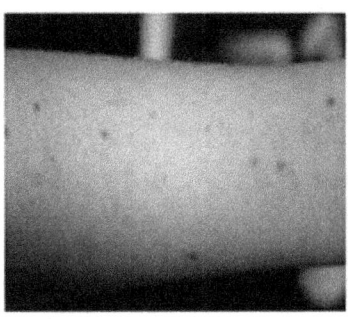

Chin Mole-- Many people have a mole on the chin. Right or left, it designates people with enviable characteristics. These are people with loving, generous dispositions. They are conscientious workers, and love to travel and acquaint themselves with the habits and customs of other peoples in distant countries. They are capable, responsible citizens, willing to accept responsibilities for family and country. [See preceding middle]

Arm Mole – Arm mole people are courteous, industrious, and have happy conjugal relations. A man may have to fight many battles if the mole is near the elbow. He may also become a widower at an early age. A woman has the same characteristics but her problems are in her occupation. [See preceding right]

Naval Mole — On a man, he will be very lucky. On a woman, she will desire to have many children.

Neck Mole — This means unexpected good fortune, if mole is on front of the neck; on either side, unreasonableness. On the back shows a need to practice frugality.

Shoulder Mole — Generally, restless, needs to travel in order to be satisfied with home surroundings. A mole on the right shoulder brings prudence, discretion; a faithful marriage partner; very industrious. On the left shoulder, this person is satisfied with any position in life, both occupational and social.

Wrist Mole — Frugal, ingenious, dependable people have wrist moles. Furthermore, on a woman shows she will have only one marriage; on a man, possibly two marriages.

Twin Mole--When there are twin moles, such as a mole on one wrist in a certain spot and an identical one on the other wrist, this is called a Gemini duality. The person possesses a dual nature. This pertains to all dual moles no matter what the location, such as legs, arms, cheeks, and so on. Two moles, side by side, are said to indicate two loves. For those with more than 2 moles, one could determine he should not have children and pass on the indications of the moleoscopy testing and the world again was saved until Physiognomy scientists came along.

Fake Physiognomy

This science without science was almost universally praised by ancient Greek mathematician, astronomer, and scientists. While you would think this stuff would not continue, by the 15th century Physiognomy science was widely accepted. English universities taught it. Around 1530, scholastic leaders started with the Greek form 'physiognomy'.

Greek Physiognomy- The first indications of a developed physiognomic theory appear in fifth century BC Athens, with the works of Zopyrus, who was said to be an expert in the art. Here are some details of Physiognomy used in Greece. *After inspecting Socrates, a physiognomist announced that he was given to intemperance, sensuality, and violent bursts of passion—which was so contrary to Socrates's image that his students accused the physiognomist of lying. Socrates put the issue to rest by saying that originally, he was given to all these vices, but had particularly strong self-discipline.*

By the fourth century BC, the philosopher Aristotle made frequent reference to theory and literature concerning the relationship of appearance to character. Aristotle was apparently receptive to such an idea, as evidenced by a passage in his "Prior Analytics"; *It is possible to infer character from features, if it is granted that the body and the soul are changed together by the natural affections: I say "natural", for though perhaps by learning music a man has made some change in his soul, this is not one of those affections natural to us; rather I refer to passions and desires when I speak of natural emotions. If then this were granted and also that for each change there is a corresponding sign, and we could state the affection and sign proper to each kind of animal, we shall be able to infer character from features.* [Remember the dog remark as we go along.]

Chinese physiognomy- This face reading, called *mianxiang*, reaches back at least to the Northern Tsung period of China. Like the others, the desire was to rid the world of inappropriate people.

18th Century Physiognomy-Another principal promoter of physiognomy in modern times was the 18th century Swiss pastor Johann Kaspar Lavater. Lavater's essays on physiognomy were first published in German in 1772 and gained great popularity. These influential essays were translated into French and English. A Belgian by the name of Paul Bouts, expanded the concepts of the early Greeks so he could make the world safer. He combined phrenology with typology (character analysis through body morphology) and graphology (character analysis through handwriting examination). He is noted for calling his three-in-one science Psychognomy. The mole thing didn't seem to track, people, again pushed for science to help them understand people they finally knew moles were just moles so someone began measuring heads outside of the bumps and nose angle. This "science" used facial characteristics to determine nature of a person. Here are some of the findings.

Round head – This shows that a person is intuitive.

Triangular head- This shows the person is impractical, but a quick thinker.

Straight eyebrows- This marks an alert and active person.

Arched eyebrows- This marks a person with a great imagination.

Large nose- This marks an aggressive person.

Narrow lips- These mark an unemotional person.

Large earlobes- These mark an independent person.

No earlobes- These mark a person who lacks sense of purpose.

High forehead- This marks an intellectual person.

Narrow forehead- This marks a great analyst.

Round eyes- These mark a naïve person.

These evaluations continued to all elements of civility. Here is a sample from James Redfield's book "Comparative Physiognomy "[1852]-*The organ of veneration is situated in the middle of the coronal region, between benevolence [Left below- hump in middle of skull] and firmness [second below- dip in the middle of the skull]. When rapt in devotional feelings, when all outward impressions are unheeded, the eyes are raised by an action neither taught nor acquired. Instinctively they bow the body*

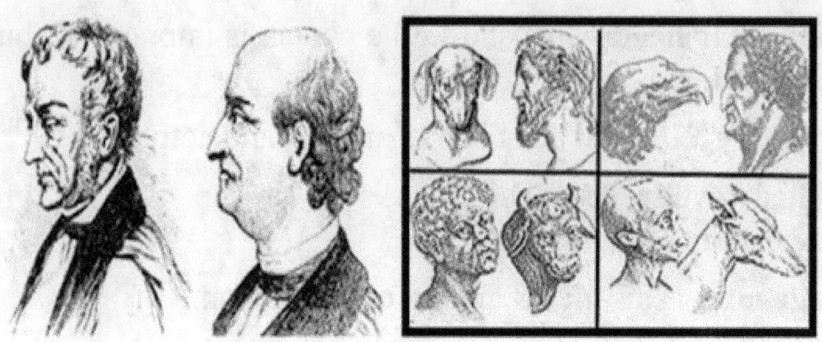

In his book Redfield "*Resemblances between Men and Animals*", he also associated people's character and appearance to animals. [Germans to Lions, Negroes to Elephants and Fishes, Chinamen to Hogs, Yankees to Bears, Jews to Goats] Here are a few of this consensus scientific disclosure. [See above right]

Fake Scientific Histories

So, we have seen consensus medicine has not helped and consensus physiology has not helped the world. Let me very briefly describe how consensus scientific history hasn't helped our world either. In the next few chapters we will look at how Consensus dictated how our history books described science and the art of invention. Rather than teaching our children the details, we took the opportunity to expand the good feeling of the greatness of the United States even if that greatness should have been shared. If history associated with science gets in the way of a particular consensus, simply change it. That's what has happened over and over and over again. In that way your quasi-science can protect the world from knowing their past. Knowing the past could change the minds of voters, citizens, government grants, Nobel Prize awards, scientific honorariums, and scientific "research" that would be debunked if the truth is allowed to be distributed. You would think it is impossible to hide the truth today with the internet and such, but it is so easy. We, as a people, want to believe what is being told to us so very badly. You probably already know some of these unfortunate lies that were established and continued by consensus but let's look at a few. This first example is an obvious one where we wanted one of our founding fathers to be hailed for another accomplishment called the Franklin stove. As we go through these, the more bizarre ones will be saved for later.

Who Made the Franklin Stove?

We finally changed the Constitution and Ben Franklin was still around inventing away. He did, indeed, design sort of a stove in 1790. It is shown below left. It was called the Pennsylvanian Fireplace, but it did not work well because he thought the smoke should come out of the bottom; see the flu opening at the base in the following example. The man who made the Franklin Stove was named David Robinson. It used to be called the Robinson Stove and somewhere along the line, as usual, history was changed and Franklin became the inventor of this practical element of many homes to make our history a consensus.

Yes; there were dozens of stove designers and manufacturers after the Franklin. What I'm trying to instill in this book is that we should be very careful trusting that what we are taught in school has much to do with truth and more to do with comfort or esoteric agenda.

1892 Who Made the First Radio?

I guess you know it wasn't Marconi, because I purposefully picked things that have been changed by history books. In this case the inventor was **Nathan Stubblefield**. He demonstrated his device over a period of 10 years to many people, but unfortunately his patent '00600457" didn't keep others from stealing his invention of 1892. Many years later Marconi became our hero.

Stubblefield's Radio Transmitter-As the father of the Radio, he said. *"I have perfected now the greatest invention the world has ever known. I've taken light from the air and the earth as I did with sound ... I want you to know about making a whole hillside blossom with light...".* After that, he locked himself in his shack and starved himself to death. His unbelievable invention completely forgotten and thrown in the trash is shown next left.

Sewing Machine Invention

Certainly, sewing would not threaten history and an accurate accounting could be written down, so who invented the sewing machine? If you said Elias Howe, you'd know who patented the idea of 1846, but many would say Isaac Singer, who was just a patent infringer from the 1850s. Both are wrong, of course. Ten years before the original patent, a man named Walter Hunt actually made one in New York, but he also wasn't the first either. It was actually invented by a Frenchman named **Barthelemy Thimmonier**. In 1831, some of these sewing machines were purchased and used by the French Army, so it wasn't like an inventor just coming up with a silly idea like many do today. There may be two reasons we don't remember him. He was French, and no one wants to admit when the French help us, and no one could spell his name so Howe and Singer became famous by consensus. Another example of this type of misrepresented invention might be the fantastic device that changed the world and allowed work to continue after dark.

Light Bulb Invention

As we are talking about consensus history you probably guessed that the scumbag named Edison wasn't the one who invented the light bulb. He was simply the first to use tungsten as the emitter and it was not considered the best lightbulb by any means. Edison is, without a doubt a bit player in this field, but he had good lawyers. In fact, there were 22 other "inventors" of the light bulb came before Edison. Here are a few.

Sir Humphrey Davy [1800] –He was an Englishman and invented the first light bulb seventy years before Edison's patent. His bulb used a Platinum emitter. Sir Humphry Davy's got a patent for his glowing carbon filament in 1800 and several others for glowing platinum, iridium, and carbon in the 1840s.

De La Rue [1820] –He was a British astronomer and chemist who created his lightbulb in 1820 by passing an electric current through a platinum coil in a vacuum tube. Despite its effectiveness, the cost of the platinum made his invention impractical for commercial use.

James Lindsay [1835] -Lindsay is said to have demonstrated an electric lamp at a public meeting in Dundee, Scotland, in 1835. However, his claims are not well documented and he did not develop the device further. His innovation used powdered charcoal between two platinum wires contained in a vacuum bulb.

Frederic De Moleyns [1841]- Englishman was granted the first patent for an incandescent bulb but his contribution was likewise eclipsed.

Henry Goebel- [1855] A controversial figure in the history of lighting, Goebel, who was originally from Germany but later moved to New York, claimed to have created working incandescent bulbs in the 1850s, prior to Edison's inventions.

Lewis Howard Latimer- [1870] The son of black slaves who fled antebellum Virginia for New Jersey, Latimer worked for another light bulb pioneer, Hiram S. Maxim, and patented a carbon filament that enabled the light bulb to burn longer. Edison eventually purchased the patent and hired Latimer in 1884. He also created the threaded socket that allows a light bulb to fit into the fixture.

Alexander Lodygin- [1874] He acquired a patent in Russia for the incandescent lightbulb in 1874. He later moved to the US and acquired patents for incandescent bulbs with different filaments. His lightbulb with a molybdenum filament and worked very well.

Henry Woodward and Matthew Evans- [1874] Woodward and Evans designed and patented an incandescent lightbulb in Canada. They attempted to commercialize their invention, but were unsuccessful, and eventually sold their patent to Edison.

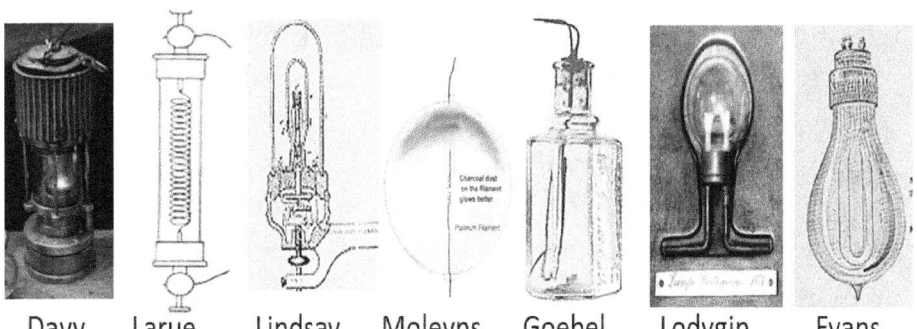

Davy Larue Lindsay Moleyns Goebel Lodygin Evans

Hiram Maxim [1875] – He developed and installed the first electric lights in a New York City in 1878. Edison had his sights out for this fantastic inventor as he had a much better understanding of patenting law. Maxim claimed that an employee of his had falsely patented the invention under his own name. All Edison had to do was prove the employee's claim was false. Rather than the patent going to Maxim, it became an unpatentable component and Edison swooped down to keep Maxim from continuing in the lamp business.

William Sawyer- [1877] He developed a lighting apparatus in 1877 and founded a company with Albon Man to produce incandescent lamps. He successfully defended his patents against the Edison company, and secured a contract with Westinghouse to use his apparatus to light the Chicago World's Fair in 1893.

Joseph Swan- [1878] He demonstrated his electric lamp to the Newcastle Chemical Society in northern England. Swan showed a workable version of this remarkable creation before Edison. He had made filaments of carbonized paper.

Thomas Edison- [1879]- Edison gained two patents for his carbon filament lamps. When Edison went commercial with his lighting system, Swan held the better patent, so Edison took him into partnership. Light-bulb performance measurements made at the 1883 Vienna Exposition of twelve inventors showed Swan's bulbs were slightly better than Edison's, but several others had, by then, outclassed them both. In 1893, lawyers for the main three lightbulb manufacturers who had not given their patents to him were sued by Edison Electric Light Company. The power of Edison and Latimer's patents and a few underhanded actions made them the instant kings.

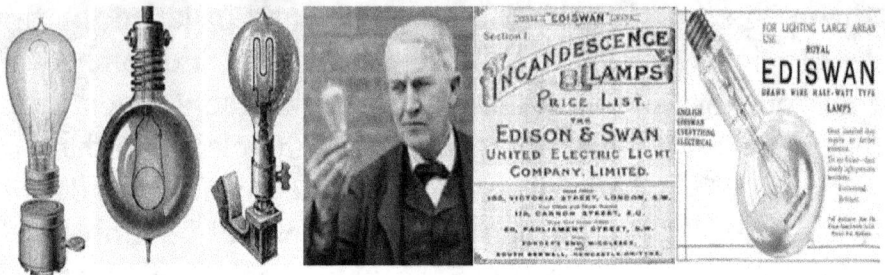

Latimer   Swan   Maxim   Edison   Eidson/Swan company and their bulb

Besides what Edison had done to the inventors of the world making light bulbs was nothing compared to how he tried his best to destroy Nikoli Tesla. Tesla was a true genius and Edison was a hack. He tried to make a DC motor and failed, but Tesla was offered $50,000 to fix it so he made it infinitely better so Edison told Tesla he was kidding and Tesla had to quit and work as a ditch digger. When he got on his feet, he found out his fluoroscope idea was now being "designed" by Edison. One made was killed and Edison almost lost an eye. Then came Tesla's AC Electric generator which was far more efficient and could be carried on wire for hundreds of miles. Edison tried to scare the public and Tesla tried to stop him at every turn. First Edison killed animals, including an elephant with AC electricity which was so horrible and then Edison made an electric chair to really show the "danger" of AC electricity and had an inmate electrocuted. Tesla's sponsor, Westinghouse, said it would have been nicer to use an axe. Tesla's went on to create radar and all types of important things and continued to give the rights to the population and to Westinghouse until Westinghouse was rich and Edison was rich and Tesla died a pauper. If you are thinking that having friends in the patent office helped Edison, let's look at the telephone.

Telephone Invention

By now you must know that the <u>consensus historians</u> are completely wrong again. It possibly should go to a German physicist named John Reis. In fact, he made one in 1861, many years before Alexander Bell became famous with his copy of Reis work. About the only change made by Bell was the use of a different voice diaphragm to increase the clarity. Our history books praise Bell. Probably German history books tell a different truth. There were telephones even earlier than Reis'.

Innocenzo Manzetti- [1849]- He considered the idea of a telephone as early as 1844, and even made one in 1864, as an enhancement to an automaton built by him in 1849. Unfortunately, lack of a patent wiped out his memory.

Antonio Meucci- [1854]- The first American demonstration of Meucci's invention took place in Staten Island, New York in 1854. An early voice communicating device was called a *telettrofono*. In 1871 Meucci filed a caveat at the US Patent Office. His caveat described his invention, but unfortunately, it did not mention the diaphragm, electromagnet, conversion of sound into electrical waves, conversion of electrical waves into sound, or other essential features of an electromagnetic telephone, so Meucci lost his place by consensus historians.

Charles Bourseul [1854]- He was born in Belgium, and grew up in France. Charles worked for the telegraph company as a civil engineer and mechanic. He made improvements to the telegraph system then Bourseul experimented with the electrical transmission of the human voice and developed an electromagnetic microphone, but his telephone receiver was initially unable to convert electric current back into clear human voice sounds. In 1854 Bourseul wrote a memorandum on the transmission of the human voice by electric currents that were first published in a Paris magazine "L'Illustration", though no known prototype was built.

The Reis Telephone [1861]- In 1860 Johann Philipp Reis was the first to produce a functioning electromagnetic device that could transmit musical notes, <u>indistinct speech,</u> and occasionally <u>distinct speech by means of electric signals</u>. Reis also introduced the term "telephon" for his device. The first sentence spoken on it was "*the horse doesn't eat cucumber salad*". In the Reis transmitter, a diaphragm was attached to a needle that pressed against a metal contact. This resembled the make-or-break design of Bourseul. The sound quality wasn't very good, but the device worked, just the same.

Elisha Gray [February 14 1876]- This inventor devised a tone telegraph in 1875. His tone telegraph, several vibrating steel reeds tuned to different frequencies interrupted the current, which at the other end of the line passed through electromagnets and vibrated matching tuned steel reeds near the electromagnet poles. Gray's 'harmonic telegraph,' with vibrating reeds, was used by the Western Union Telegraph Company. On February 14, 1876, at the US Patent Office, Gray's lawyer filed a patent caveat for a telephone on the very same day that Bell's lawyer filed Bell's patent application for a telephone. The water transmitter described in Gray's caveat was strikingly similar to the experimental telephone transmitter tested by Bell about 1 month later.

Alexander Bell [February 14, 1876] Bell was the 2nd or third to obtain a patent for an "apparatus for transmitting vocal or other sounds telegraphically", after experimenting with many primitive sound transmitters and receivers. The big difference was <u>Bell was an astute and articulate businessman with influential and wealthy friends</u>. All of this helped him win in patent court against other claims.

In keeping with the train of thought let me help you understand what is hidden in history book concerning Movies.

Battery Invention

Some of us remember that Benjamin Franklin first coined the term "battery" to describe an array of charged glass plates and that Alessandro Volta invented the voltaic pile and discovered a practical method of generating electricity. Constructed of alternating discs of zinc and copper with pieces of cardboard soaked in brine between the metals, the voltic pile produced electrical current. He invented his battery in 1800 and we named the measurement of electricity "volt" to honor him, but many previous discoveries are ignored by consensus scientists as they don't match up with what we want to think and the evidence seems to make many feel less in control.

60000 BC-Geode Battery- In New Mexico was found a piece of an electrical connector of some kind underlined embedded in rock. X-Rays showed the pins were made of steel. Besides that, a battery was found inside a California Geode as shown below. I placed a common "D-cell" next to it for comparison. These would have been in used over 60 thousand years ago.

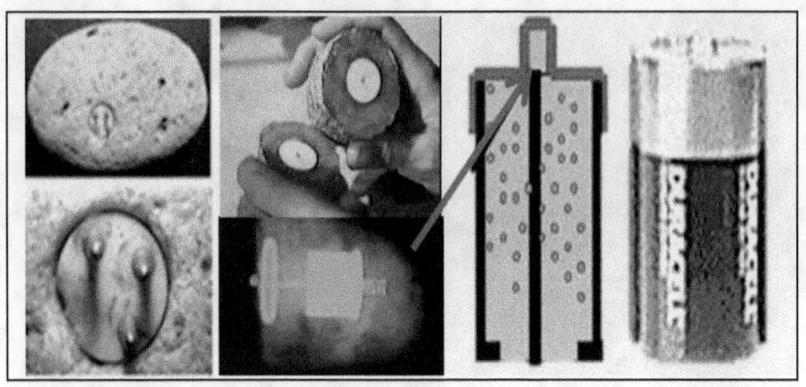

4000BC- Baghdad Battery- Then scientists found over a dozen ancient "copper-iron" batteries in Iran from about 6 thousand years ago. They would have provided about 1.2 Volts each for causing filaments to glow or electroplating [bottom row left].

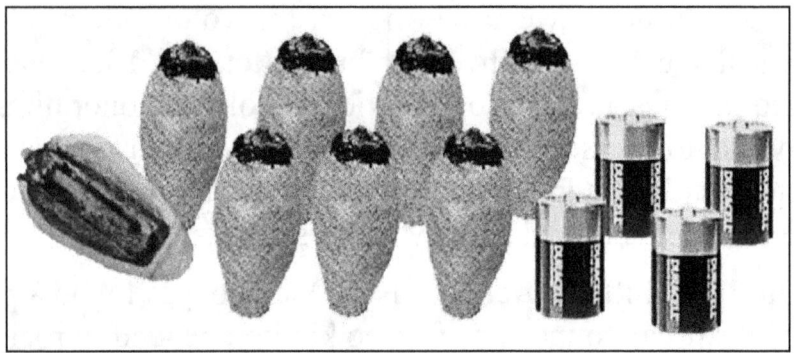

3000BC- Indian Battery-Then they found Sanskrit instructions in India for combining many "copper-zinc" batteries together to get 110VDC. The mixture was really zinc powder embedded in mercury making contact with a copper plate all in copper sulfate. With that combination the maximum voltage that can be obtained is only 1.5 Volts, so many "batteries" had to be connected as shown so that Indian electroplaters could do their work. The group below would only provide 13.5Volts.

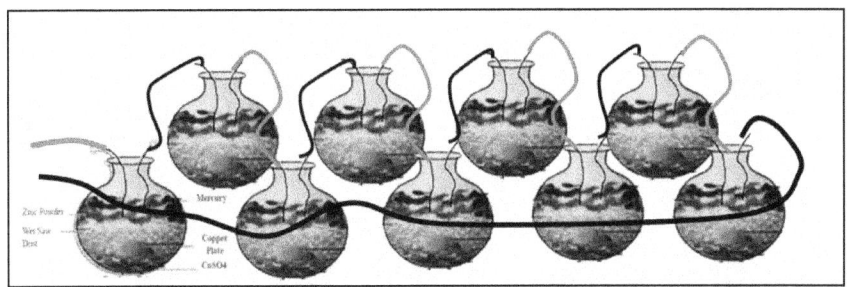

Today, we do the same thing. Hopefully, you can tell that electricity was in use thousands of years ago and continued to be used. In my fantastic wisdom I made this assumption. If batteries were used around the world, someone may have known about electricity.

Nuclear Invention

Ok! We all know this one as the United Stated built its nuclear bomb and hit Hiroshima and Nagasaki to end what has been called World War II. Our German and American scientists worked day and night to make the deadly thing, but there is a problem with us claiming such a thing just to make us feel better. We completely ignore the evidence of a more devastating event many years before we reinvented atomic bombs.

3000BC-Radioactive skeletal remains of people who died suddenly in the ancient city called Mohen jo Daro, Pakistan and hundreds of "huddled" human-remains found in the now deserted streets.

3000BC- a massive sea of glass in the middle of the Libyan Desert as if a massive explosion not relating to a meteor somehow turned miles of sand into glass. The map shows the massive explosion point and one of the millions of pieces of melted desert sand produced.

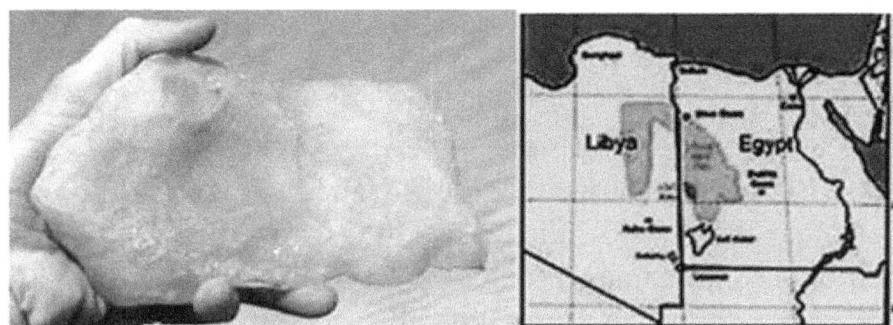

3000BC- Melted walls and pottery in Pakistan, Scotland, France, India, U.S.A, and other places have been found, as if a weapon with unbelievable heat, fused the stones together. Below are the remains of clay pots in Mohen jo Daro, Pakistan.

3000 BC- Enough processed nuclear material was determined to be missing from the 16 ancient nuclear plants located in the Oklo Mountains of Gabon, Africa to power New York for a year. Mysteriously, <u>tons of the processed Plutonium are missing</u> and have been missing for thousands of years. The image shows location of the various processing area.

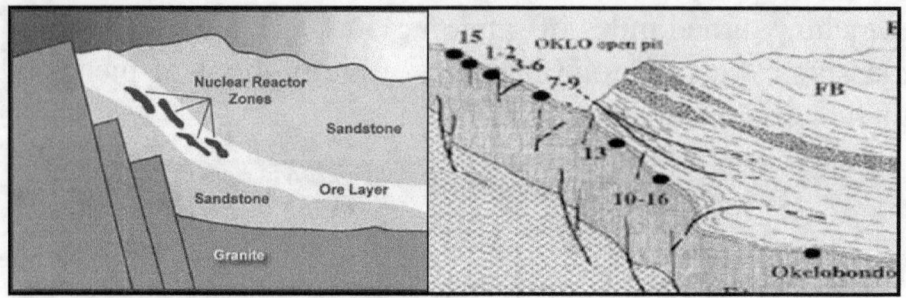

3000 BC- A well-protected, multi-walled, circular building in Peru that looks very much like a nuclear generating plant is still extant. The images below are of the 16-foot triple wall protection area in Sacsayhauman, Bolivia around the mysterious triple walled circular enclosure similar to the nuclear processing plant in Russia shown to the right.

3000 BC- Thousands of people rushed to live underground in Peru, North America, Turkey, Malta, China, India, Scotland, and around the world, as if trying to protect themselves from nuclear fallout. Some bomb shelter cities were over 20 stories deep underground. The following graphic shows some of the underground fallout shelters found at Skara Brae, Scotland.

3000BC- fortifications for above ground dwellings used 2-foot-thick walls to try to keep death out. The examples below are from some of the dwellings on the surface at Malta.

3000BC- Almost 50% of all major DNA mutations of humans occurred at this time as if some DNA modifying radiation, somehow, affected almost all the people around the world.

3000BC—Hundreds of miles of wide ancient roads in the Uranium rich Northwestern section of New Mexico. In our distant past, it must have been even more richly saturated and, apparently, still has the remains of ancient ore extraction roads while the actual machinery has long since disappeared. The map following shows a small section of these roads in New Mexico near the Colorado, Utah, Arizona borders. The roads are shown as heavy lines on the diagram. In the center of the map a section has been blown up to allow more details of the hundreds of 30-foot-wide, very flat, specialized roads which makes you wonder.

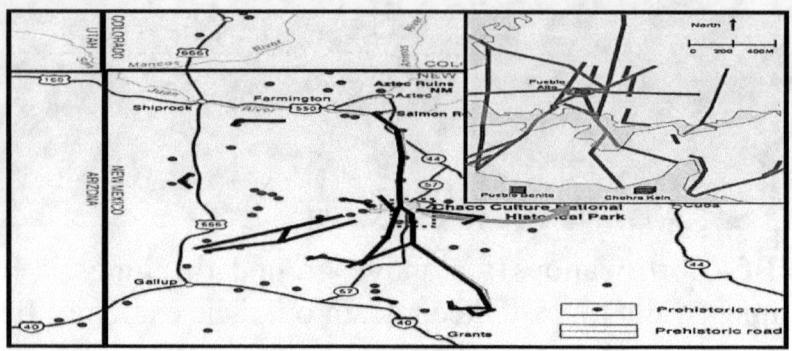

3000BC- U.S.A. *"The Pioneer"* -According to this 19th century newspaper, researchers found near the Missouri River, *"-an old cemetery of fully 100 acres in extent filled with bones of a giant race. This vast city of the dead lies just east of the Fort Lincoln road. The ground has the appearance of having been filled with trenches piled full of dead bodies, both man and beast—who appeared to have lived in a high state of civilization. This had, evidently, been a grand battlefield, where thousands of men ... had fallen."* The giant bodies, huge quantity of dead and burying animals and humans together, all, point to this same war where killing thousands at a time was possible.

3000BC- *Jasher 9:35*-36 *The outcome was divided in thirds. Those who said, we will ascend to heaven and serve our gods, became like apes and elephants. Those who said, "We will smite the heaven with arrows", the Lord killed them, by making them kill each other. And those who were left -became scattered upon the face of the whole earth.* This Biblical text describes only a third unaffected by the horrors of the war and millions dying thanks to some pretty horrible weaponry.

3000BC- Indian *"Rig Veda"*- *The Arya go on from fight to fight, destroying city after city; thou hast <u>destroyed the hundred Cities of Vangrida</u>; <u>Thunder-armed</u>! breakest down the seven Walls. The thunderous weapon did make thy missile. <u>Armed with his missile he wandered and shattering the cities of the Dravidians</u>. Arya <u>overthrew a hundred cities of stone</u>. The thirty thousand Dravidians died from magic power and weapons.*

3000BC-Indian *"Mahabharata"* - *Billowing smoke clouds like <u>giant parasols—contaminated food- people's hair fell out</u>—It <u>could burn 50,000 men to ashes in seconds</u>. <u>Flying spears could ruin cities</u> full of forts "All points of the compass were lost in the <u>darkness</u>. <u>Fierce winds began to blow</u>. <u>Clouds roared upward, showering dust and gravel</u>. The very elements seemed disturbed. The sun seemed to waiver in the heavens. The <u>Earth shook, scorched by the violent heat</u> of this weapon. <u>Elephants burst into flames</u> and ran to and fro in a frenzy, over a vast area, other animals crumpled to the ground and died. <u>Animals were burnt pools</u> and lakes began to boil".*

Some would say this description mimics that seen in Atomic Bomb tests and explosions.

3000BC- "Ramayana"-<u>*Charged missiles*</u> *mingled with each other and were surrounded by <u>fiery arrows that covered the Earth</u> and heaven. There was increased burning. - All were scorched by missiles; they felt the <u>fire that burns the world.</u>*

3100BC- Egyptians called the end of this great war "Zep-Tepi" [New beginning], Mongulala called it the "end of the Blood Age", Indians called it the "New Age of Kali". The PreMaya initiated "a brand-new calendar" with 3100BC being the starting date as if whatever happened was so horrifying that instant relief of its ending initiated a "new Life". Consensus scientists ignore it all.

Movie Misunderstanding

There is consensus that Thomas Edison made the first movies but that ignores all evidence. It just makes Edison look better. If we want to know who made the first movie, it will be hard to determine as we have found them in the very ancient world.

First Movie 6000 BC-What is known as the Burnt City, Zabol, of Persia [Iran] gives us a great glimpse into the world of movies. The movie machine is shown below. The antelope, when viewed on the bowl-shaped screen, as it is turned, jumps up to eat food over and over. I have blocked out 6 consecutive screens showing the antelope jump. It was a short movie, but no one had to pay the actors.

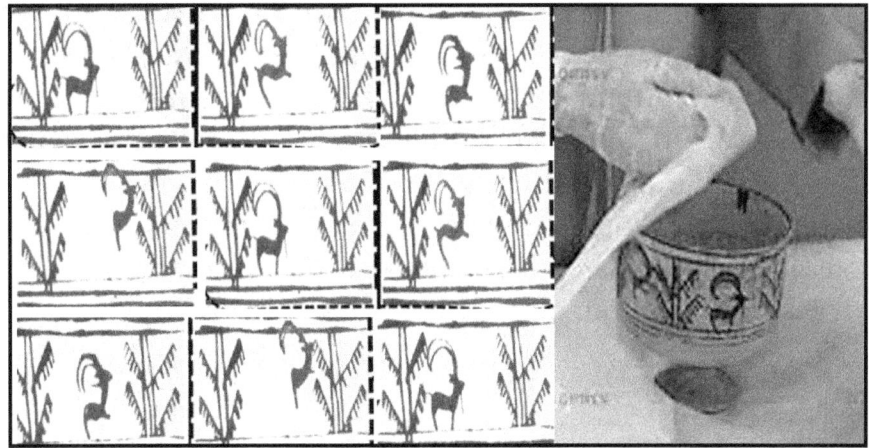

Other sites apparently had similar animals they wished to view. The upper left was the same Persian movie machine while the 2 immediately below were found in Mohen jo Daro [Ancient Pakistan] and the one on the right came from Egypt.

The images following include a couple of the many combs found, patterned dishware, and a couple of many gam

Phenakistoscope [1832] –In 1832 a new type of movie system called the phenakistoscope, built on what the movies were like 8 thousand years ago, was re-invented. This was before the invention of photography. A variety of optical toys exploited this effect by mounting successive phase drawings of things in motion on the face of a twirling disk just like the movies made during ancient times. A direct view and projector variant are shown next along with a few of the movie disks of dancing, fighting, skeleton walking, and acrobatics.

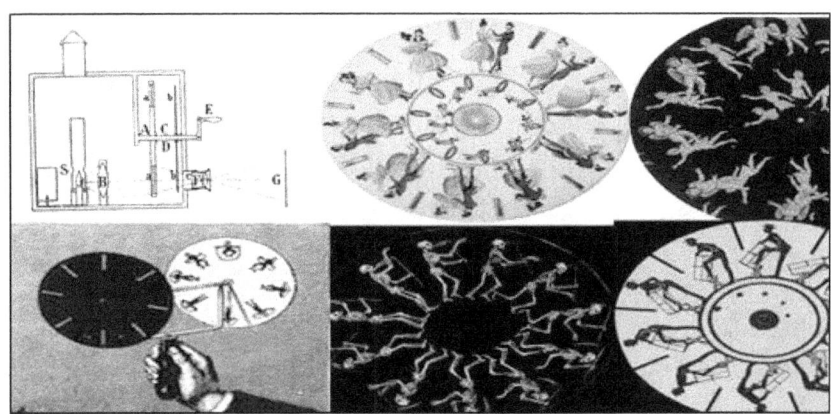

Zoetrope [1834] - Sometimes the images were placed inside a rotating drum so the movie machine had a different name. The Zoetrope machine sometimes was powered by steam as shown below and small slits in the drum allowed the action to be witnessed as shown below.

Daguerreotypy [1839] Louis-Jacques-Mandé Daguerre, perfected the positive photographic process known as daguerreotypy which added to the realism.

The first live action Motion picture [1877] This required many camera-trippings in succession as a horse passed in front of the camera. Once the images were pieced together, a live horse showed up in a motion picture machine. This was accomplished by the Englishman, Edward Buybridge. He also used a flashing light to make the images become alive without blinking. As shown next, besides horse galloping he also did naked women walking down stairs and naked man walking, along with many more similar movies.

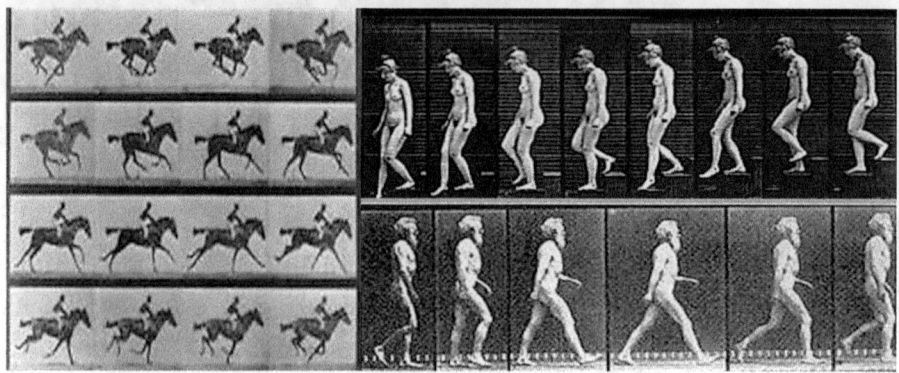

I was told, his galloping horse movie was very popular.

One camera Motion picture [1882] The French physiologist Étienne-Jules Marey, took the first series photographs with a single instrument in 1882; once again the impetus was the analysis of motion too rapid for perception by the human eye. Marey invented the chronophotographic gun, a camera shaped like a rifle that recorded 12 successive photographs per second, in order to study the movement of birds in flight. These images were imprinted on a rotating glass plate and later ones were put on, paper roll film. Marey subsequently attempted to project them without much success. Those who came after would return their discoveries to the realm of normal human vision and exploit them for profit. Some of his great movies included naked man riding a bike, naked man running, Man in tights pole vaulting, and naked woman walking.

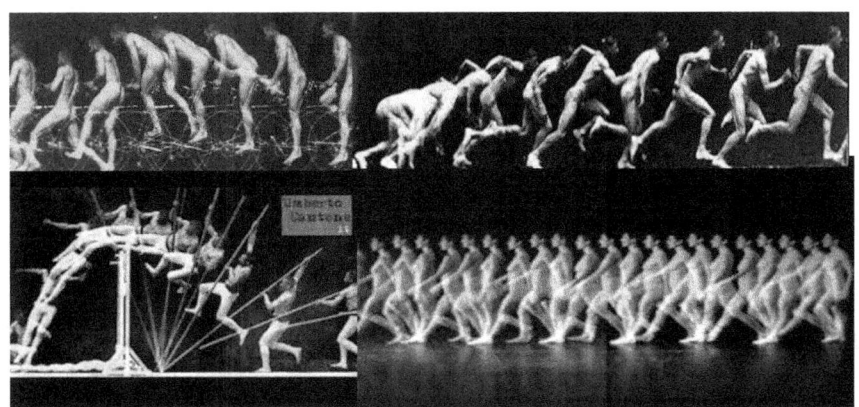

Chronophotography [1887]- Hannibal Goodwin developed the idea of using celluloid as a base for photographic emulsions so movies could be made longer. The inventor and George Eastman made the first celluloid roll for making films in 1889.

Kinetograph Motion Picture Machine [1888] Within a few years the first innovation to use the celluloid came from William Dickson in the West Orange, N.J., laboratories of the Edison Company. The Kinetograph is shown next.

Edison and the Lumière brothers [1888] –Using the Kinescope film, the only thing Edison invented was a way to synchronize a light and the image together for easier viewing.

He perforated the film. Some of the initial movies are shown below.

The movie was watched in a device similar to the Zoetrope peepshow system, but a light bulb would light up each time the image passed by as shown next. It was allowed to be longer than the old Zoetrope projector. The device was called the Kinetoscope. Starting in 1894, Kinetoscopes were marketed. The images below show the device and a Kinetoscope parlor.

Don't get me wrong here. Edison should absolutely be praised for putting those little holes in the film, but we also need to recognize that Iranian that started the whole movie business 8 thousand years before Edison put in his holes.

That brings us to flying. I assume everyone knows about the development of flying, but are we really told the truth?

Hidden Flying

Possibly, the worst tragedy of consensus rewriting of scientific history is with flying. According to many books even today, the Wright Brothers hold claim on the first powered flight. Besides being a total lie, it is worse by being heralded as a reason for American greatness. I'm not saying the Wright brothers were great engineers and helped us reestablish flight by their actions. I'm just saying we need to be told about the total picture so we can work from the truth. For this anomaly we need to go way back in time.

Ignored Bible References

Our surviving Biblical texts are full of details about this mode of travel. One extremely detailed account in the Bible talks about 4 flying ships [wheels] seen at the same time. In that instance [Ezekiel 10], the "ships" were accompanied by 4 "fiery thrones". Many people have heard of this episode, but there are many more. Let's look at both the thrones and the wheels mentioned in ancient Hebrew texts in some detail. Modern consensus scientists simply say the Bible history is bunk. I suppose looking at dinosaur bones they could claim them as bunk as well.

Thrones and Fiery Chariots of Biblical times-The flying fiery chariots were all similar in that they were fiery, fast, and fearsome. The early Jewish people called these chariots,

Merkaba, but while a name was given to them, they were forbidden to speak of them, so you have to wonder why the objects had a name. The early Jewish people might not have said the word merkaba, but they did draw them and write about them.

If you're interested in these flying ships, at least read one of the ancient Biblical books named "Enoch". Notice that, in the book of Enoch, it clearly indicates that the ships went into space and that a huge ship was in orbit around the earth. Weird isn't it? We're trying to do that now.

2 Kings 2:11- *And it came to pass, as they still went on, and talked, that, behold, there appeared a chariot of fire, and horses of fire, and parted them both asunder;*

Enoch 74:15- *I saw likewise the chariots of heaven, running in the world above to the gates in which the stars turn, which never set. One of these is greater than all which goes around the world.* [This ship going around the world sounds similar to what we do today without space ships, doesn't it?]

"The Divine Throne- Chariot"-For more detail, we need to look at the Book called *"The Divine Throne-Chariot"*. The most detailed reference of a flying ship in ancient times comes from this particular book that was found among the Dead Sea Scrolls. Here is an excerpt from the ancient book. From this account, there can be little doubt that people witnessed flying ships after the flood. Here are just a few examples.

Bless the image of the Throne-Chariot above the firmament, and - praise the majesty of the <u>fiery firmament beneath</u> the seat of his glory. [Flames came out the bottom of the throne-chariot just like a rocket.]

And between the turning wheels, angels of holiness come and go, [Angels ride inside the turning wheel.]

as it were a fiery vision -about them flow seeming rivulets of fire, like gleaming bronze, a radiance of many gorgeous colors, of marvelous pigments magnificently mingled. [The flames shoot out in brilliant colors.]

Ascending they rise marvelously; settling, they stay still. [It could hover and/or ascend in the air. This hovering thing is very interesting.]

Ignored Very Ancient Flights

Babylonian Flying machines-The images below show a Babylonian depiction of Merkaba. This one also shows a man driving the air-ship with the same type of round portal. By the way, for those thinking these are people riding inside birds, there are no birds with portals and no heads. Many carvings were made with a driver in a NON-BIRD like the one shown. Several images have multiple passengers.

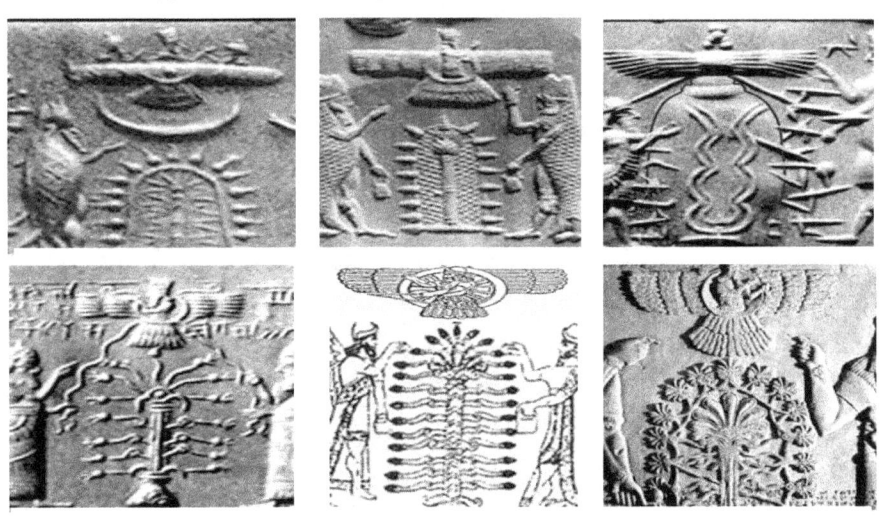

Everywhere you turn you find these flying machines were COMMON around 4 thousand years ago or before.

Besides winged aircraft, we find all types of rockets. The one in the middle looks very much like or Space-shuttles except they were driven before the massive nuclear events of 3000 BC.

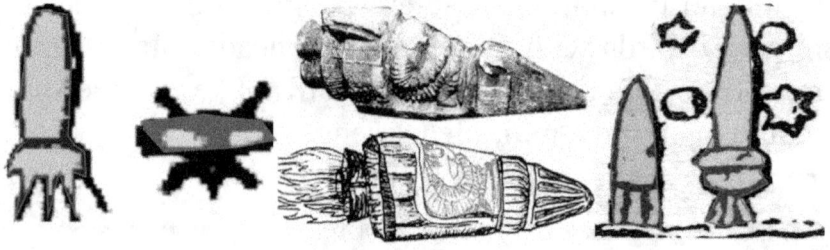

Pakistan Flying-While India and Pakistan are really one and the same, this little section is really about one of the cities in the northern portion of India. For this evidence we go to the ancient city of Mohen-jo-Daro. Here we find drawings depicting the ruling "goddess" who was so strong she could fight two tigers at the same time. She was shown with a flying ship over her head. [See next left] Some have tried to identify it with something else, but none have made any sense. The second image is another flying disk image found in India.

Australian Carvings-In Australia, the aborigines drew flying ships on the cliffs. Some looked like they had landed, while others were shown over water and even windows are shown. The drawings below right are believed to be at least 6 thousand years old. [See previous right]

France and Spain Flying-Cave dwellers were depicting these flying ships over **40 thousand years ago**. Below are samples from French, Spanish, Chinese, and Japanese Caves.

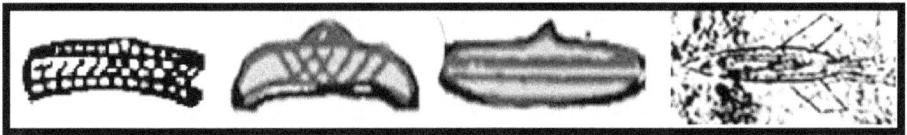

Central American Flight-Hundreds of indications of flight are easily found in ancient American documentation, and physical evidence can also be found. Here are just a few indications form the Maya "Codex Nattal", the Olmec documentation, and Columbian models.

*Codex Nattal-*In the Nattal, one of the pictorial images clearly shows what is believed to be a rocket ship with thruster engines and flames. [See below left along with 2 suits for flying where there is no air.]

Olmec Rockets- The Olmec carved this representation of a rocket-like object with the body of one of their gods at its base. This could have been a headdress if the person at the base was very, very strong, but I doubt it [See below 2nd]. The third representation is thought to be a missile piercing a mountain or some type of pointed object tunneling through a mountain, but to me it looks like a rocket that has been pierced. [See below third.]

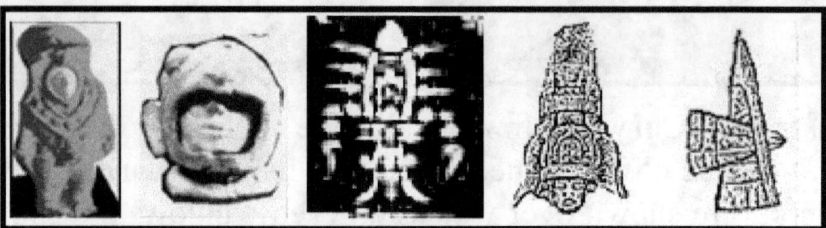

South American Evidence- Nine of the Columbian golden jet models manufactured about 2 thousand years ago are shown above right. They appear to be more like the jets of today which shows that the Americans had a wide assortment of air transportation possibilities. If the Americans had all of these contraptions in the not too distant past, where are they today? Certainly, the metals would have disintegrated, but what happened to the knowledge about building them and are the ships we see today the newer versions of those used in the past? I think the answer might very well be yes. To the right is a huge desert in Nazca, filled with landing strip details and images of animals that can only be seen from an airplane.

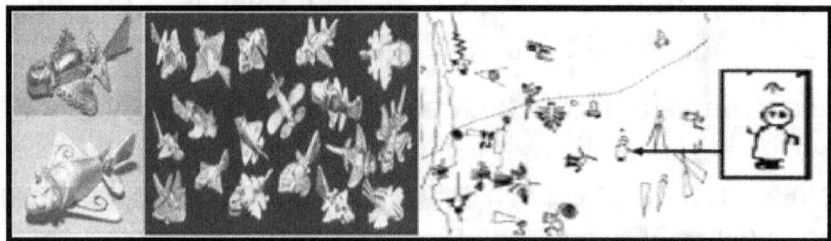

Oriental Flight- Bronze airplanes models have been discovered in China, too. Bronze models were found in ruins in Liaoning from 2500 years ago. While some try to say these are fish images, the tail gives them away. They are clearly jet planes. Like the South American ones, these objects seem to be beyond the imagination of people in olden times. We can presume that they were not beyond their imagination because they saw them in the air. [Next left] Another image was found on a cave wall [5th image following] and three more planes models were found similar to the jet models. [Next right]

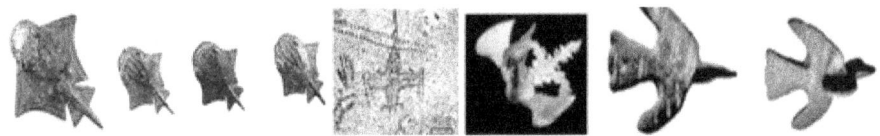

2200BC, Chinese used flying machines- Emperor Shun reportedly not only *"made a flying ship, but also a parachute of sorts,"* 4200 years ago.

2000BC, Chinese built some- Emperor Tang wrote about a *"flying chariot that was built and flown. The craft was destroyed to keep the secrets away from outsiders."*

300BC, Chinese saw flying machines- Chu Yuan wrote of a *"flight taken in a jade chariot at a high altitude over the Gobi Desert"*. He described how the *"ship was not affected by wind and how an aerial survey was performed, 2300 years ago"*.

200BC, Chinese saw some more- The historian, Ko-Hung, wrote about a flying craft with rotating blades to set the machine in motion [helicopter?]-2200 years ago.

French Flying-Below left is an image found in Niaux near the Pyrenees, France from over 3 thousand years ago. In the middle is another image showing a sky battle and a pilot wearing a space suit of some kind from Peche Merle Cave, France.

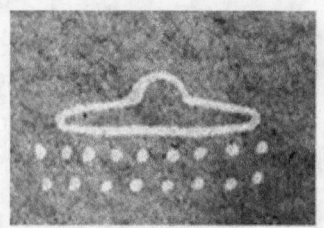  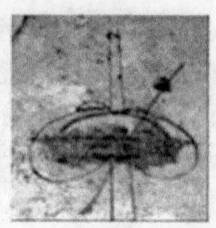

Italian Flyting-The last image was found in a cave near Pompeii, Italy that was estimated to be 5000 years old. Notice that the ship has a 3 person crew.

Greeks Flying in Space

*Greek Legend-*In discussions about battles between the gods we find the following: *"Hot vapor lapped the titans, flames unspeakable rose bright to the upper air [outer space], lightning blinded their eyes."*

Apparently lightning weapons were used in outer space.

Pilots from Nepal-In Nepal, an indication of a strange visitors was found. A carved plate has been found which was manufactured around 4 thousand years ago. The carvings on the plate shows a large headed being similar to that depicted as one of the UFO pilots currently seen an elliptical shaped object above him, which is very similar to reported UFOs of today. That is odd enough, but what I want you to see here is that it looks like the Big-headed guy is traveling from Mars down to Earth in this football shaped thing. There also is a monkey looking thing coming from where Venus would be and a Lizard thing possibly leaving Mars, but let's mostly look at the Martian.

 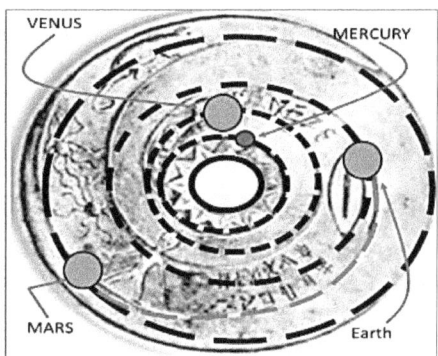

Egyptian Description-1500 BC- As recorded, The Pharaoh Thutmose III saw *"silent, foul-smelling circles of fire and flying discs in the sky"*.

Macedonians saw them 329 BC - Alexander the Great told of *"2 great silver shields, spitting fire around the rims that dived repeatedly at his army as they were attempting crossing the Jaxartes River"*.

Romans Saw them-399 BC - Julius Obsequens wrote in "Prodigia" about a number of Roman sightings, *"-- a round object, like a globe, a round or circular shield, took its path in the sky from west to east."*, another section reads, *"-- there came a terrific noise in the sky, and a globe of fire appeared burning in the north. In the territory of Spoletum, a globe of fire, of golden color, fell to the earth gyrating. It then seemed to increase in size, rose from the earth and ascended into the sky, where it obscured the sun with its brilliance. It revolved toward the eastern quadrant of the sky."*

Painter's Provided Documentation -15th Century- So far, we have seen testimony of sightings and pictures and carvings of not completely humanlike humanoids, but there is even more evidence that some strange flying was going on even through the Middle Ages. People didn't have cameras in those old days, but they may have had something almost as good. They had painters. Fantastic painters that saw UFOs just like many have done today. The collage below seems pretty much like what has been reported today as unidentified flying objects, but these portions of paintings came from the Middle Ages.

The two on the bottom right clearly show human passengers were believed to have been in these strange ships. As shown below, many saw these "ships" so clearly, they could make out features including portals; glowing beams; and saucer shapes just like our present-day phenomenon.

During the Middle Ages people kept on seeing strange flying machines in the sky just like we do today, and as I showed, the pilots have been seen, are humanoid and somewhat similar to us.

Australian Flying-The same construction was noted in Australia in the rock etching shown to the left from over 8 thousand years ago. The flying saucer manufacturers seemed to like the three-orb concept and kept it for a long time.

Indian Flying-The picture to the right is a more dramatic picture of what is typically called a UFO. The three orbs at its base are identical to other images and detail that many believe show it is of a similar design to that used by the ancient VIMANAs that fought in a Bharata War 3100BC according to a large number of Indian Texts.

More Modern, Pre-Wright-Brothers Flights

Let's say we discount the hundreds and hundreds of indications of flight thousands of years ago, we still have a consensus issue as the best one could say is that the Wright brothers were the third or so to prove powered fight of a heavier than air craft in the United States, but people loved the wright brothers story so they became famous.

December 1903-We are all taught that the Wright Brothers invented the airplane and made the first successful powered and manned flight at Kitty Hawk in 1903 and the flight lasted 1,500 feet. [Below left]

August 18 1903 – This was the first flight of Jatho Zweidecker's airplane. It flew 18 meters. A second flight was done in November 1903. The second flight went 60 meters at a height of 2.5 meters. Both flights beat the Wright Brothers. His airplane is shown previous right.

1899- While Augustus Herring who designed and built bicycles in Ohio just like the Wright Brothers, indicated he also had longer flights, on October 11, 1899, Herring successfully flew along the sandy shores of Lake Michigan at Silver Beach in St. Joseph, Michigan. This particular historic flight carried him approximately 50 feet. Eleven days later, he repeated the trip with a 73-foot flight, as was witnessed by a newspaper reporter: [Next left]

 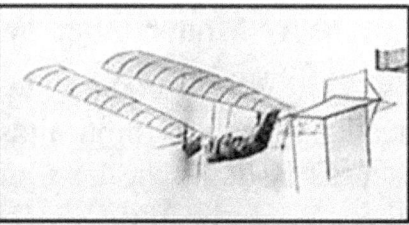

1896- Before Augustus was a somewhat famous individual named Alexander Graham Bell – his contraption flew up to heights of about 100 feet in the air and traveled about 2500 feet before the steam engine ran out of fuel. [Above right]

1869- Steam powered lighter than air craft Avitor Hermes successfully flew the powered airplane, shown next left. It was assisted by gas and it didn't happen in the United States so the fantastic event was ignored. That still is not far enough back in time, but this Avitor character was still a great inventor.

1864- Steam powered Aircraft- In France, Felix du Temple powered an aircraft that reportedly flew as early as 1864. Its design is shown above right and his amazing flight seems to have been ignored.

1844 Steam powered Aircraft- A man named William Henson built a steam powered craft that flew without a passenger. [See next left]

1848- Steam powered Aircraft- A man named Springfellow designed a Steam powered aircraft that successfully flew for a about 120 feet. His vehicle is shown above right.

Of course, many flew contraptions powered by gas or hot air so those should be considered powered craft as well and before that hundreds of gliders flew all over the place.

1888- Lighter than air craft- Campbell Air Ship, powered by an Edison electric motor, its 18,000 cu. ft. envelope supplied by Carl E. Meyers, and built a cost of $2500 by the Novelty Air Ship Company of Brooklyn, N.Y., for Professor

Peter C. Campbell; the first flight of which was made December 8, 1888 from Coney Island to Sheepshead Bay, piloted by Carl's wife. [See next left]

1851- Lighter than air craft- Petin's Aerial Navigation System. This was actually 4 dirigibles integrated with a steam engine. [previous right]

1647-An Italian inventor called Burattini designed a flying dragon for the King of Poland. His prototype was able to lift a cat into the air and he was paid by the King to develop a full-sized aircraft. The dragon had four pairs of wings – two middle pairs to provide lifting surfaces (showing an understanding of the concept of lift in aerodynamics), a pair to the rear to produce lift and forward movement and a pair at the front for forward motion only. [See below right]

1500- Back to the early 16th century and Leonardo da Vinci gets you closer. This creator invented his form of airplane and helicopter around 1500. [See helicopter below middle]

850- A Muslim doctor named Abbas Ibn Firnas, jumped off a wall in Cordoba, Spain wearing a suit of feathers and bird wings strapped to his arms and legs. When he jumped from the wall, he was able to glide a while and even, according to observers, ascend a little. He found it difficult to land because he did not understand the way in which a tail operates to stabilize flight, but we still seem to forget him; possibly because he was Muslim.

700-Kites may have been used for human flight in China and Japan from around the 7th century. Marco Polo reported seeing men flying in kites and Japanese folk stories refer to kites being used to allow men to fly and there may have been laws outlawing this practice.

While I understand how people can be enamored by charisma and charm, it is simply not right to base science and history of feelings of national pride or the idea of promoting an ideology rather than the science itself.

Now let's talk about the really old indications of intelligent aeronautics from thousands of years ago. This is ignored totally because of fear in jeopardizing the concept of evolution and the master race of giant people who ruled the entire world thousands of years ago. Genesis simply called them the giants of old and the Bible called the Anakim. The Sumerians changed the name to Annunaki and the Mongulala of South Americans changed the name to Akamim. In Egypt they were called the "Lords of Amenti" while the people of India called them the Araya. In North America these same giant rulers were called the Archaics and everyplace you go these ancient people used air travel. There is little probability these three were different groups and there is little probability that they didn't have the technology of flying in what the

Indians called Vimanas. After the Bharata War [3400-3100BC] the world was devastated. It was recorded that 1/3 of the population of the entire world was dead and almost 50% of all major human DNA mutation occurred during the war according to Haplotype scientists. One of the mutations was a major loss in brain function. While the evidence of these things is all around us, many are fearful of the ramifications of these events and make up their own history by consensus.

Fake Science Quips

***Speed Aerodynamics*-** It was also "proven by consensus" that if a train goes faster than 21 MPH, the air will be sucked out and everyone will be suffocated. Some smuck went 22 and the science community simply disregarded their last acclamation.

***Geography*-**From the 16th century, European experts in geography "knew" that California was an island separate from the North American mainland. Maps of the time show a large island on the left of the land mass and California continued to appear this way even into the 18th century.

***Geography of the Bible*-**There was at the time also a knowledge that California was an earthly paradise like the Garden of Eden or Atlantis. The matter was finally put to rest indisputably on the 1774-1776 expeditions of Juan Bautista de Anza. Interestingly, it is likely that within 25 million years, Baja California and part of Southern California really will separate from North America due to tectonic plate movement.

***Astronomy*-**Astronomers with all their mathematical background told us no stones can fall from the sky, because there are no stones in the sky. Soon someone decided to test the material in a planet or star and some said oops!

***Astronomic Flat Earth*-** The truism that we had a flat earth was common until the 3rd century BC. Eratosthenes calculated the circumference of the Earth without leaving Egypt. Eratosthenes knew that at local noon on the summer solstice in the Ancient Egyptian city of Aswan on the Tropic of Cancer, the Sun would appear at the zenith, directly overhead. He dug a hole and measured the length of the shadow. The shadow length indicated the Earth was a sphere and was 24,860 miles in diameter. Of course, Columbus knew just like all ship captains knew he would never reach China when he left Portugal as it would be over 15 thousand miles away. He was a liar.

***Astronomic Geocentricity*-** Geocentricity was the truism that the earth was the center of the Universe and that all other objects move around it. The view was universally embraced in Ancient Greece and very similar ideas were held in Ancient China. The idea was supported by the fact that the sun, stars, and planets appear to revolve around Earth, and the physical perception that the Earth is stable and not moving. This was combined with the belief that the earth was a sphere. The geocentric model was eventually displaced with the work of Copernicus, Galileo, and Kepler in the 16th Century.

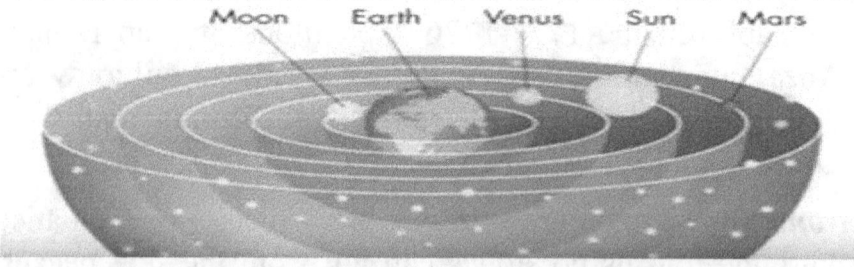

The previous image of a geocentristic universe not only shows the whole idea of Geocentricity initiated in the time of the ancient Greek [300 BC], it also shows that by this time the whole flat earth theories were all taking a back seat.

Abiogenesis- Anaximander, a Greek philosopher who taught Pythagoras, believed that at some point in history, humans and animals had been <u>born from the soil</u> spontaneously in adult form; otherwise they could never have survived. Many Scientists right up to the 19th century knew this, and some even wrote recipe books for making animals. One such recipe, to make a scorpion, calls for basil, placed between two bricks and left in sunlight. This consensus fact was not finally put to rest until 1859, when Louis Pasteur sort of proved it wrong once and for all.

I put this chapter in to get you ready for one that has been in the news quite a bit lately called climatology.

Fake Climatology

We have all been hearing it. Global warming, Global Warming! Quick get rid of you underarm spray as it might let hydrocarbons into the air! Even with 50% of our Electrical energy coming from coal, we need to eliminate the use of burning coal or all our lands will dry up! Now build electric cars to eliminate the destructive use of gasoline. Wait a minute! I'm back to coal again making electricity for the things. All Greenhouse gases except for the only one that could possibly cause a problem became the villains of the 21st century as they pushed us closer to destruction. H_2O gas was ignored even when everyone knew it causes so much temperature change because <u>no one could tax it</u> or <u>make money form demonizing water vapor</u>. You also hear almost all scientists agree that human caused global warming will destroy the earth when 30 thousand scientists in one event discredited the claim only to be threatened by the consensus crowd.

While many indicate they are climatologists, the word is nebulous and easily mischaracterized. There are climatologist claims from individuals with degrees in mechanical engineering, mathematics, meteorology, oceanography, physics, chemistry, geography, geology, astrophysics, statistics, astronomy, atmospheric sciences, electronic and industrial engineering, earth sciences, environmental sciences, and more specific fields within those already mentioned. Scientists like Al Gore, with his global warming degree in Psychology, told us *the Earth is dying because of man* and we gave him the Nobel Peace Prize even though he only brought loss-of-peace, an abundance-of-fear, and made a $billion by buying quasi-green industries and proclaiming the end of the world. He went to accept the prize in his massive, private, fossil-fuel-guzzling jet, and praised how we were turning a corner with horribly inefficient windmills, minimally effective solar panels, making plants artificially grow faster with DNA modification to reduce the need for fertilizers, and on and on. This is such an important issue; we must make a determination of truth. Unfortunately, there are many levels of truth in a world of what we can call vain truth. Many times, consensus scientists HELP their own "vain-truth" look like reality.

We know something is happening, but what it is and how it will affect us needs to be understood without shouting unfounded claims to support some pet project or money scheme. While our temperature has not been affected, greenhouse gas concentrations SEEM to be slowly going up for the last 200 years as shown next. While some tell you reasons, no one actually knows, for sure, why it is happening and what effect, if any, this increase will have. It should be noted that while these increases have been steady, the gases identified here make a **tiny** portion of our atmosphere. Also notice the chart to the right shows CO_2 levels increase AFTER the temperature rise, not before.

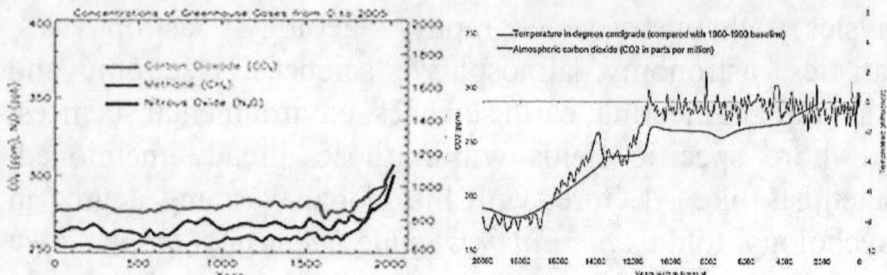

Antarctica Ice Loss Lie-Even the government paid group that is supposed to protect us where climate is involved got into the changing the data to force fear. The first graph shows NOAA data to push fear with an apparent massive loss in Antarctic ice, but they <u>also published the second one</u> showing the Antarctic ice is slowly increasing and never claimed it meant anything. Of course, the first one has been manipulated by showing a very small snippet and selecting a single point on Antarctica that had this temporary reduction.

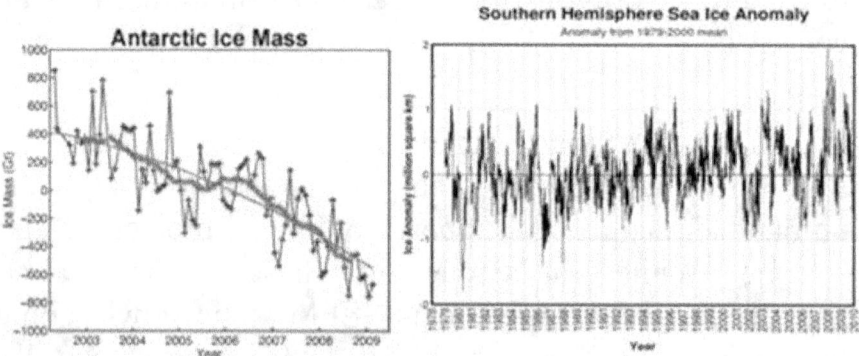

Sea Surface Temperature Lie-Here is another alarming bit of data passed on by those with ulterior motives. The first shows NOAA <u>calculated</u> destruction of our oceans as the seas overheat and the water levels rise. Let's look at a second set of data from satellites. The Sea surface temperature as measured by NASA satellite which shows <u>a general downward slope</u>, but that isn't all. The average from 1998 till now has been drifting downwards and since 2009 the temperature has been declining continuously. One of the charts is a lie and I'm not sure satellites can lie effectively, but psychology major climatologists might have a motive to add fear.

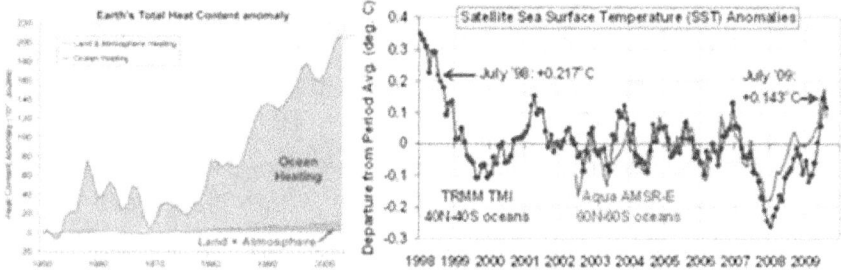

CO$_2$ levels in the Ice Core Lie- The next one is really sick so go throw-up first before reading the data. Our great friends at NOAA and IPCC took the thousands of years of Ice Core CO$_2$ levels and added the atmospheric CO$_2$ levels from Hawaii to make it look like the re was death from people burning coal. The problem is that airborne CO$_2$ doesn't fall to the ground so it could not be captured. First look at the scare graph and then I have other data. The second graph shows where the Ice Core ended and how the airborne data was spliced.

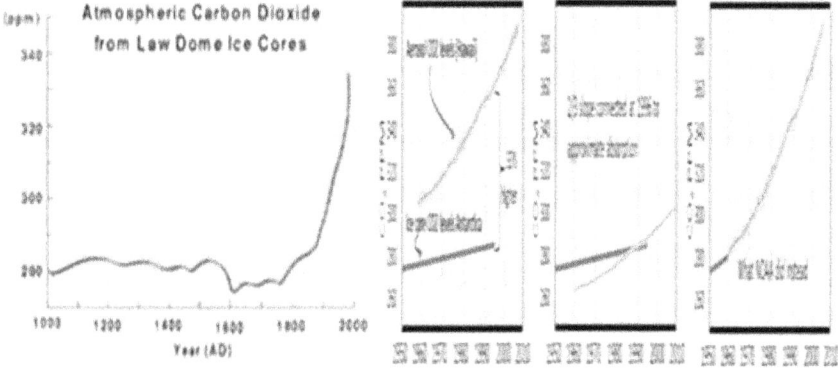

IPCC Data Tells Us the Treachery-Luckily, you don't need to take my word for anything. The IPCC told on themselves in a paper presented in 2009 that tried to confirm massive increases in CO_2, but what they ended up showing was TREACHERY. Here are some snippets. I have not put them together just to make it look bad, but I didn't think you would want to go through all the mess so I boiled it down a little. The IPCC acknowledged CO_2 has something called a short residence time, stating:

> *"The turnover time of CO_2 in the atmosphere is about 4 years. This means that on average it takes only a few years before a CO_2 molecule in the atmosphere **is taken up by plants or dissolved in the ocean**.*

As you read through this remember the Hawaiian Aerosol CO_2 detection is the thing that is causing the entire ruckus. The data taken included substantial amounts of CO_2 that would be absorbed into plants and never get into the Antarctic Ice.

> *"The CO_2 response function used in this report is based on the revised version of the Bern carbon-cycle model used in Chapter 10 of this report. About 50% of a CO_2 increase will be removed from the atmosphere within 30 years, a further 30% will be removed within a few centuries. The remaining 20% may stay in the atmosphere for thousands of years".*

CO_2 Absorption lie- Besides all that calculation of how long CO_2 stays in the air, it can't absorb in the air easily. Those suggesting CO_2 is a major greenhouse gas, seemed to have never even looked at an atmospheric absorption map as shown next. Water is by far [Over 99% of absorbed energy] the most absorbed large molecule gas that can cause horror or blessing. It is very difficult to absorb CO_2 into air. Even in areas where CO_2 can absorb, water is already so absorbed that it still has an inability to be introduced in our atmosphere. Here is what we can suppose about the large increase in CO_2. There has been a tiny reduction in water vapor which shows up as a much higher CO_2 percentage. No matter what, we must understand that water vapor controls our atmospheric temperature----not CO_2. CO_2 has only 3 small wavelengths of absorption while water is absorbed in huge amounts. The second darken area in the graph is the only absorption wavelength of CO_2 while the top curve is the total amount of energy in the atmosphere. CO_2 cannot absorb into the Atmosphere. Change must be caused by something else.

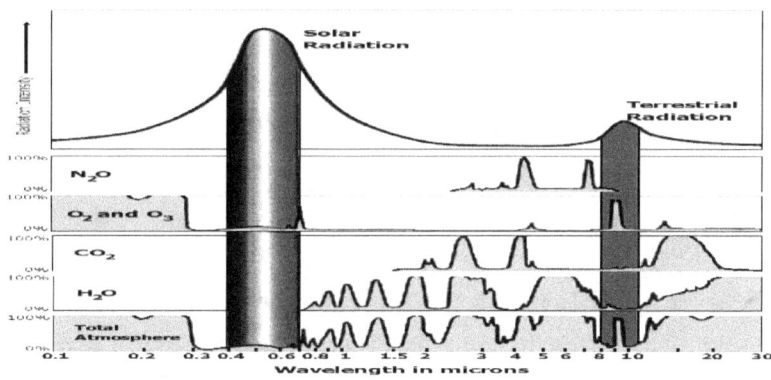

A crying shame- Some of these fake scientists have speculated that when the CO_2 absorption gets to 560 ppm there will be a substantial temperature rise that could be devastating. [Oops! I laughed a little writing that one.] A massive rise in CO_2 already noted has driven our temperature to the unbelievable increase of about 0.3 degrees making a further increase to 560 ppm from the current 390 have a total increase of less than 0.5 degrees. We simply are not affected by CO_2 in the atmosphere at all. 560 ppm (the dreaded doubling), temperatures should rise by *another* 0.2 to 0.5°C *ONLY*. IPCC [NATOs Global Warming gurus] estimate of 2.0 to 6.0°C, and this totally unfounded and without scientific merit.

Not much CO2 in the First Place-If CO_2 in the atmosphere has risen from about 290 to 390 ppm in just over 100 years, which has been recognized, and if only 5% of all the greenhouse gases are man-made, then we can conclude that only 5% of the extra 100 ppm could have been caused by mankind. This is only 5 ppm over 100 years so we can say the following:

The other change is due to the 95 ppm from **naturally** *produced CO_2.*

Even if CO2 had ANY affect almost none is by MAN-Added to this, IPCC claimed that we will get 3-5°C more warming in the next 50 or so years. Mankind would only contribute 5 to 10% of that. That is only 0.2 to 0.5°C, even if we keep using coal. Just think about it man-used CO_2 makes up only 5% of the 5% of the 1% of the atmosphere or 0.0025%. The following chart represents one view that is much less than 5% for man uses and remember CO_2 makes up the tiniest fraction of the atmospheric Greenhouse gases, so divide that tiny piece of the pie down much farther and see if it can have ANY effect on weather.

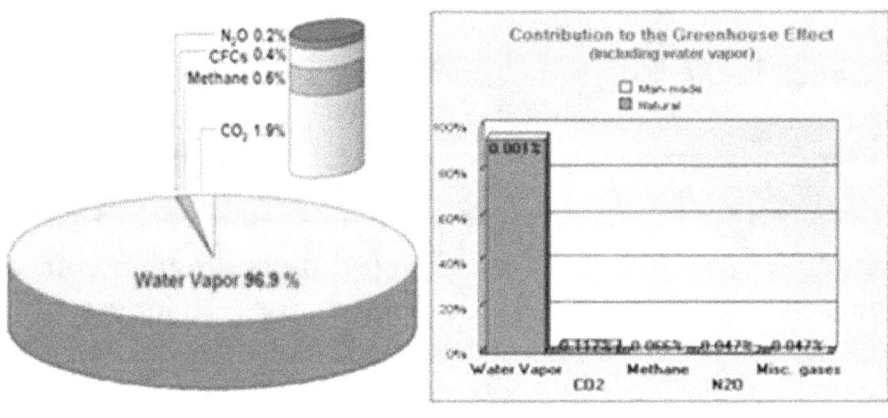

CO_2 and Temperature

Massive temperature changes occurred well before man knew about coal- Let's hypothetically say CO_2 is changing our temperature. The charts below show both CO_2 levels and temperature captured in the Ice Cores from Greenland and Antarctica. The one on the left is from Antarctica over the last 50 thousand years that I showed a little while ago. The thin line that begins below the erratic temperature curve shows something interesting. CO_2 doesn't change until after temperature changes as temperature controls CO_2 rather than the other way around. The second graph is from Greenland in case we didn't see things right, we see that Temperature, the erratic line changes well before CO_2.

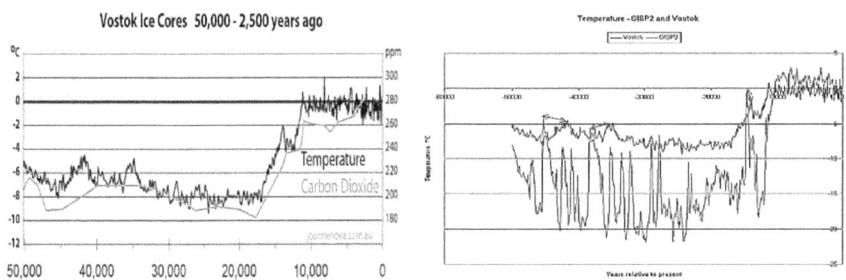

The second thing to notice is that the worse thermal change recorded in the last 15 thousand years happened 10000 years ago and very few automobiles were even on the roads.

Temperature increase makes CO_2 not the other way around

Let me let you in on a secret. The carbon dust coming out of smokestacks has nothing to do with colorless, odorless CO_2. CO_2 is made by animal respiration including people, but mostly by plankton and sea animals and it needs trees to convert the CO_2 to more Oxygen so we won't run out. Cut down the trees to allow biofuels to be made [a supposedly save the Earth] and you are going to increase the amount of CO_2 because there are no trees. This is what is happening in Europe.

Destruction because of the Lie-Because of all of the fear tactics of these consensus scientists, today Europe has mowed down massive forests to produce biofuels so that they can say they are not killing the earth and without trees to absorb CO_2 they are increasing the issue and killing the Earth [but they seem to feel better because consensus scientists are lying to them]. Bio-fuels from grain are not helping the situation. Besides the obvious loss of CO2 eaters [trees], use of "grain-based fuels" will greatly increase food prices and roughly 30 million people are expected to be severely deprived. The USA will use up to 30% of the annual corn crop for alcohol production for vehicles alone. Ethanol production requires energy as well to make it economical which is generally done with those nasty hydrocarbon methods. The actual cost/gallon is much the same as other liquid fuels, but the miles per gallon consumed by vehicles are much lower than gasoline. One estimate is that one tank full of ethanol for an SUV is obtained from enough corn to feed one African for a year. Of course, it isn't working by itself so worldwide ethanol plant subsidies in 2008 alone totaled more than $15 billion to reduce the food to feed the people of the planet to supposedly save them.

Ocean Boiling Lie-The huge increase in CO2 levels shown as a dotted line had absolutely NO effect on the Oceans that have stayed the same temperature. James Hansen's belief of CO_2 caused global warming is not supported by the tropic's data in the least nor is his crazy prediction of boiling oceans. <u>You can keep driving your gasoline car and the earth will not even know it</u>

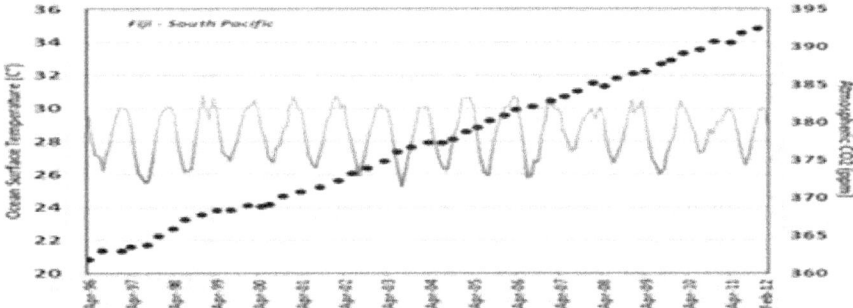

Temperature Trend Lie- If we expand out to the last 4000 years, the Greenland Ice core shows temperatures mostly stayed the same the whole time except for the little dip we started coming out of about 300 years ago. Since that time, <u>the temperature has been trying to get back to NORMAL</u> and the consensus scientist KNOWS IT!

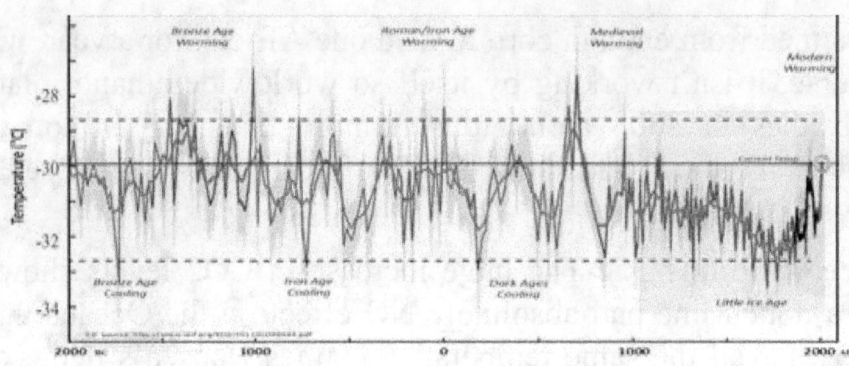

ARGOS Data Treachery-Scientists designed 3000 ARGO buoys that just float around and they took temperature measurements since 2003. These buoys show absolutely NO thermal increase [but the published data from NOAA somehow showed massive changes that the buoys "SOMEHOW" missed. The NOAA team decided that the information from them should not be used. One reason noted is that they weren't floating near the Arctic. Some might wonder why NOAA would have helped place these things and later decide they were stupid. Some begin to suspect that there was a large network of politicians, corporations, and scientists that were conspiring to promote the fear of "global warming".. <u>despite evidence clearly stating no such "global warming" exists</u>. With only $22 billion being pushed into the global warming epidemic, you might wonder why some would try such a scare tactic.

NOAA Published Truth then pulled it to publish a Lie-The National Climate Data Center and National Oceanographic and Atmospheric Administration [NOAA] put out a chart showing there was no significant temperature rise in the United States from 1940 until 2010 and that the spring was the coldest in the 115 year record, but at the same time they told everyone to ignore this data and focus on eliminating the Coal Industry.

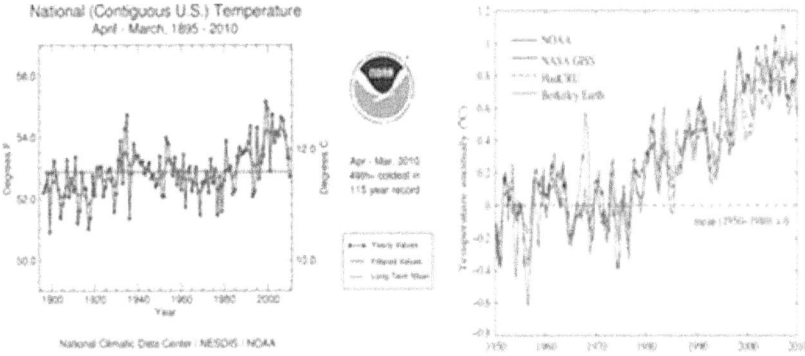

After realizing their mistake they put out a completely different chart so that people would fund their pet projects associated with Earth annihilation. By focusing on the slight rise since about 1960, the danger begins to look real.

17 Year Cooling Disregarding Lie-NOAA got fancy with this one. According to NASA's own data, the world has warmed 0.36 degrees Fahrenheit over the last 35 years, starting at the fairly cold year-1979. Even this would show a massive increase of 0.1 degree per year over that short time. The NASA Remote Sensing Systems data also shows that since 1998, the average temperatures around the world have been steadily decreasing as shown below.

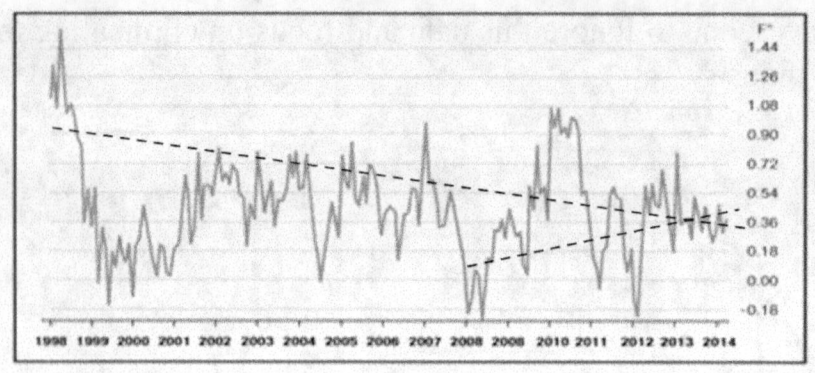

Source: NASA, NOAA and Remote Sensing Systems

According to this graph, the world is 1.08 degrees cooler than it was in 1998. NOAA has not used this data when "Informing the World" about the condition of our Earth. To make it look like horror, just truncate until a rise is shown [2008 until today shows a 0.32-degree rise]. This is not the new data I mentioned to correct the previous longer-term graph. That is next.

Just Plain Lying-NOAAs current US graph is shown below left [same as before]. <u>*Now we know it is all a lie*</u>. Note that there is a discontinuity at 1998, which doesn't look right. <u>Globally, temperatures plummeted in 1999-2000</u>, but they didn't in the US graph. Note that measured data below right shows that by 2008, temperatures were back down to the 1989 level. But in the NCDC data, 2008 is half a degree warmer than 1989 making the temperature LOOK like a disaster when there is almost no change at all. Please note that the faked chart to the left has been used to justify an enormous number of "charts showing the destruction of our world.

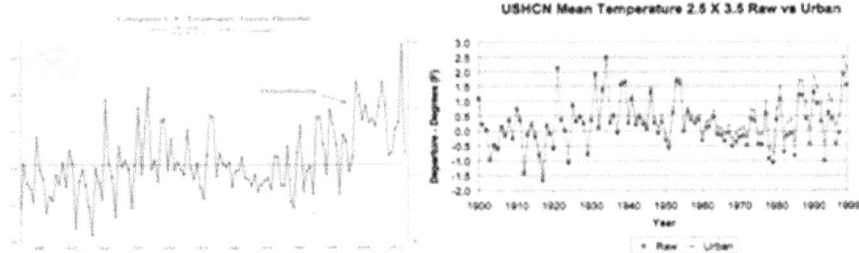

By putting the 2 together we can easily see the treachery. The top graph is from RAW data and the bottom one is the "doctored" chart making everyone want to give money to Green Technologies to protect them from this FAKE temperature rise.

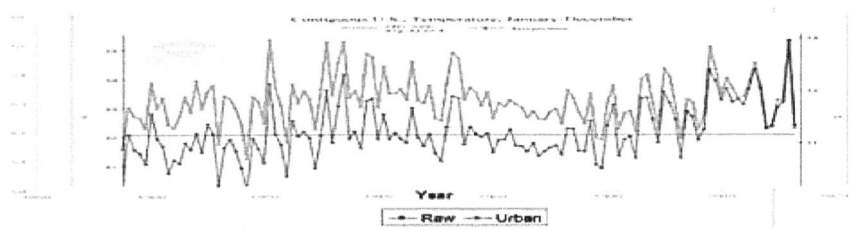

Top line is truth. Bottom line is the MANIPULATED NOAAs US temperature record is completely broken, and meaningless.

Adjustments that used to go flat after 1990 now go up exponentially. Documented Positive Adjustments are implemented as negative to amplify fear for monetary gain.

Just erase the high temperatures of 1940-The best way to describe the subterfuge is to talk about the 1940s heat wave. This was a huge thermal "spike" occurring as the Earth recovered from the mini-Ice Age. If you were somewhat devious, you would like that spike to go away as it shows the temperature today is not significantly different than 1940 levels. This can be done 2 ways. Show the thermal rise since

1965. This is the favorite one, but you can also change the 1940 peaks to smooth out "anomalies that don't go along with the "THREAT". In this case it seems unthinkable, but this is another way to make people buy Solar Cells for their homes; <u>especially when the US Government pays for most of the installations to eliminate the use of Coal that we have mountains of, is the lowest cost energy producer, and assures we can stay energy independent.</u> The following shows the actual temperatures recorded and the one without the 1940s data is what you see on TV.

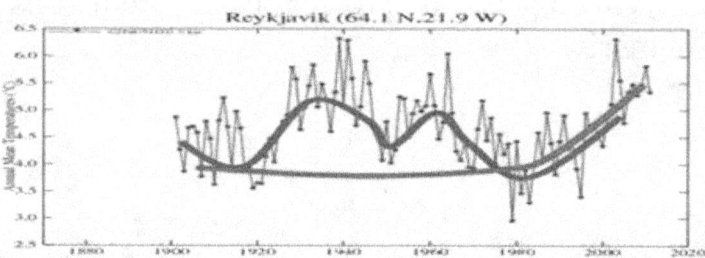

After the practice was revealed from some cleaver undercoverwork and retrieval of internal EMAIL traffic, we have started to see a few of the practitioners coming clean about their part in the underhanded fear mongering.

"Ice" Lie-The Nobel Peace Prize, get rid of coal, billionaire stated the following, *"The North Polar ice cap is falling off a cliff. It could be completely gone in summer in as little as seven years. Seven years from now.* "The images below show the Ice cap in 1990, 2007, 2013 and 2015. From the minor reduction in 2007, the Ice caps have been almost steadily increasing to over 63% larger in 2015.

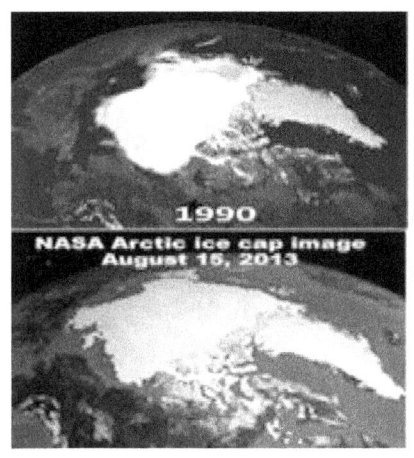
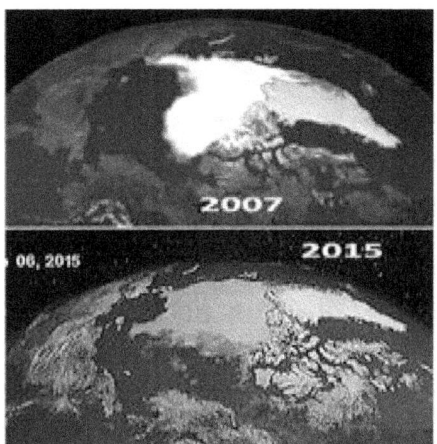

"Man Caused Death" Lie-Al Gore claimed *CO_2 emissions from Human factories were destroying our world as massive amounts of factory effluent were making the Earth's temperature go out of control and <u>97% of scientists agree it's real</u>, it's man-made, and it's dangerous.* Certainly, he knew the satellite data and all the rest, but he was making a fortune. He also knew many scientists were begging for people to listen to them as Mr. Gore misrepresented everything.

The Sun not CO_2-All those greenhouse gas things are the real reason for temperature fluctuations. The real culprit is the sun. All of the heat in our atmosphere is solar heat. If the sun gets brighter and burns away clouds, things happen and when the sun generates cosmic and X-Rays that hit the earth other things happen. Instead of just shining, the sun blast energy out in spurts. If we look at the correlation between temperature and total solar irradiance, we see a much better relationship and we can begin to understand CO_2 is not the main player. In fact; the tiny CO2 gas levels could not possibly affect weather no matter how many gas guzzling cars there were! It is the activity of the sun (sun spots, <u>solar flares</u>, modification of other galactic cosmic radiation from outer space, the effects of solar wind, and magnetic flux), that affects the radiation arriving on earth. Here is a big one. The sun moderates cloud cover! Approximately 1% of the atmosphere is greenhouse gas and 90-95% of that is water. $\underline{CO_2\text{ is about }0.05\%}$ of the atmosphere. But only 5% of that 0.05% is man-made!

At 1.3 Billion times as large as the Earth, the sun makes 99.9999% of all the energy locked in our atmosphere. Some even suggest is gets colder at night and warmer in the daytime when the sun heats the air. From the next graph we can see that from 1978 the temperature is getting slightly warmer in parts of Greenland, so what did we do differently before 1978. The graph shows that there has been no appreciable slowdown in the increase of CO_2 in the atmosphere, but there was a fairly significant reduction in the temperature between 1940 and 1978. Sunspots control our heating if we want to limit heating, eliminating CO2 emission will have NO EFFECT.

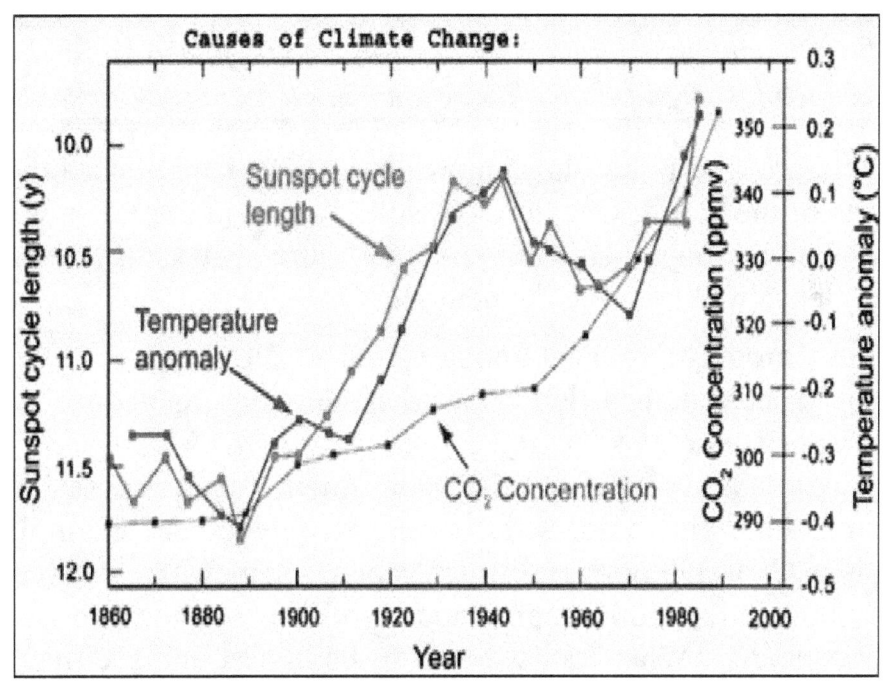

Climate Control

Attempts to correlate solar activity with global temperature have been going on for some time now as you would expect. Measurements from the SORCE's *Spectral Irradiance Monitor*s show that solar UV variability produces colder winters in the US and southern Europe and warmer winters in Canada and northern Europe during solar minima. Here is what we are finding.

Cosmic Clouds Issue-Solar wind-mediated galactic cosmic ray changes, seems to affect cloud cover which would change the thermal characteristics. Much of the solar effects have been clouded by the evils of NOAA trying to keep their favorite stocks going up, but I think you can see climate is almost totally controlled outside our atmosphere. The cosmic ray change over the cycle changes clouds. Let me be direct here!

All environmental scientists know this and they also know that the most significant changes are in the polar regions.

Therefore, the poles show more thermal variance than other parts of the world. Get rid of Cosmic Rays and the Poles will stay more regulated, but making sure cows have less flatulence will not save the penguins.

Most Dangerous Greenhouse Gas-The IPCC can't see this simply because they don't consider the main greenhouse gas as water vapor. If <u>95% of the greenhouse gas is water</u> and it is; we must attribute clouds as one of the major causes for temperature increase. Let me give you an example. With CO_2 at 390 ppm and water vapor is somewhere between 10,000 and 40,000 ppm and wreaking havoc on our atmosphere. When can we start to be honest and see that if we have been in cooler periods, it is natural to expect that it will get warmer so quickly, there must be a much larger driving force than a trace of CO_2 gas in the atmosphere? The Sun is a major player [man is NOT]. **<u>Cosmic rays from interstellar space modified by the sun's solar wind</u>** plus the sun's and earth's magnetic fields are next largest factors after the sun itself. Following this is CLOUDS. Way below all these is what we call non-water greenhouse gases which play a very minor role. Here is something the NOAA didn't tell you.

If the CO_2 is removed from the atmosphere, water vapor will absorb that band of infrared energy and make just as much heat as if the CO_2 had never been there.

A ray of hope- In June 2016, North Carolina lawmakers approved a bill (HB 819) requiring the state's Coastal

Resources Commission to plan only for sea levels rising at a rate based on historical records. In other words, coastal planners are banned from considering models that might produce accelerating sea level rise because of things like rising global temperatures and melting polar ice caps, especially considering none of the predictions have come close to being correct for all of the reasons I have stated. They are all based on lies. Sea levels have risen less than 10 inches along the North Carolina in the past century; consensus quasi-scientists had suggested sea levels might rise <u>by as much as 39 inches</u> before 2100, but Americans seem to be waking up and ignoring the travesty called climatology just like Phenology, Craniometry, Aryanism, and all the rest are being ignored.

Error of Nuclear Timers

What I'm talking about here is nuclear decay as a form of "accurate dating. Initially nuclear decay was tested to see how consistent it was and it seemed pretty accurate so everyone got on the bandwagon of nuclear decay. Consensus scientists could just look at percentages of nuclear isotope in a substance and instantly know how old it was. We could even time the earth with something called Argon testing.

Oh! Happy days! Oh! happy days!

For some time now, ALL have known the huge issues in this previously established timing baseline. You have been told dinosaurs died 65-million-years-ago and the beginning of the Mesozoic Era was 300-Million-years-ago. You were told, over and over and over again. They even proved to you that was truth by telling you lead, potassium, and even carbon isotopes decayed at a set rate; just see how much of an early isotope is left and read the date. Besides, the dinosaurs are buried underground and turned into stone so there had to be a long time for that to happen.

Ancient Earth consensus-While the earth is ancient, it is definitely not as old as has been told to you. Many of geologists today still tell you that radiometric dating has narrowed the age of Earth to about 4.5-billion-years, give or take a couple of percent. We now know that the dating method is inaccurate and scientists not pursuing that vain truth I talked about earlier are refining the timing more and more each day. The Earth and everything in it are much younger and so are the characteristic stabilities of the planets in our Solar System. Researchers at Purdue and Stanford have found evidence that-

*- **radioactive decay rates are not constant at all.** Don't tell our children, don't modify any science books, don't make statements on Television, don't completely retime the earth and all of the various ages we were taught in school. That would disrupt our history, science, paleontology, evolution, Astronomy, and just about everything. It would be much better to leave it alone. [the consensus quasi-scientists said]*

'On December 13, 2006, a magnificent solar flare flung radiation and solar particles toward Earth. Measuring the decay rate of manganese-54 during the flare proved to be very interesting as the decay rate dropped during the time of the radiation fallout. It was determined that solar neutrinos zipped through space and affected Mn-54's decay rates used in the experiment. Just think about this. They were testing a single solar flare event and the change was significant. The sun has these things all the time. It was also found that the decay rates of silicon-32 and radium-226 showed seasonal variation, according to data collected at Brookhaven National Laboratory on Long Island and the Federal Physical and Technical Institute in Germany. This error was just the material sitting there with almost no outside interference. Wood buried in igneous rock in Queensland Australia has been dated to 40 thousand years, while the basalt around it dated to 45 million years. Both dating subjects should have given the same date, since the igneous rock was formed at the same time the wood was buried. Many of the "data-ologists" don't tell you about major errors like this.

Lava Errors- Excess argon-36 was found in three out of 26 lava flows in recent times. So, <u>Argon/argon testing would show a much older date that actually was "KNOWN"</u>. This is believed to be because there was too much of the argon-36 in the first place. In the Grand Canyon lava flow testing showed lower levels of lava was younger than the top layers. At different volcano sites, that had eruption in 1949, 1954 and 1975. The same thing was noted. Geochron Laboratories of Cambridge, Massachusetts dated these samples. Even though the <u>oldest of these samples are just over sixty-years old</u>, the lab tests provided ages that <u>ranged from 270,000 years to 3.5 million years old</u>. Additionally, we go to Mt. St. Helens and its eruptions in the 1980's. Samples there gave old ages in the range of <u>300,000 to 2.7 million</u> years. Hopefully, you are beginning to see that we know less about how old we are than you believed before reading this. If neutrinos from a single solar flare can make things look older, what if the entire Earth was closer to the sun? I know that sounds odd, so just keep it in the back of your mind right now as we try to find some standard for dating.

Nuclear Decay a Bad Timing Method

Today we know that the nuclear decay dating of things including Electron Spin Dating and Uranium Dating, Thorium Protactinium Dating, Oxygen Sediment Dating, Lead-lead-lead Dating, and Argon Dating [which we originally used to date the ages of the Earth] are terribly flawed. The old standard carbon 14 dating also seemed in jeopardy. Dating beyond about 30 thousand years was <u>much younger</u> than tested. If there had been nuclear events [bombs or even volcanic eruptions] the apparent timing was changed drastically. Other methods had to be employed to determine how everything should be timed, but classroom information was not changed. That would confuse the students. I'm going to prove to you how you have been lied to by consensus quasi-scientists. This will give you a better understanding of the lengths some will go to when they believe something, no matter what the evidence shows.

Standard Geological Timeline

Era/Period/Epoch	Time (M yrs. ago)	Time (T yrs. ago)
Archaeozoic Period	5000-1500	50,000-3000
Proterozoic Period	1500-545	3000-1000
Cambrian period	550-500	1000-900
Ordovician period	500-440	900-800
Silurian period	440-410	800-700
Devonian period	410-365	700-600
Carboniferous	365-300	600-500
Permian period	300-250	500-400
Triassic period	250-212	400-300
Jurassic period	212-145	300-200
Cretaceous period	145-65	200-100
Tertiary period	65-1.8	100-40
Pleistocene period	1.8-0.01	40-10
Holocene period	0.01-0	10-00

The middle listing of dates above is the "STANDARD" that had been presented in our classrooms, while the last column shows a somewhat closer, more accurate time line that has been verified by MANY, MANY non-nuclear decay methods. Even with the mountain of evidence showing how nuclear decay cannot be used, the middle timing is still heralded as the master in many schools and books being used to teach our children <u>without basis just like most of the other stuff we have been talking about</u>. I know it is difficult to believe historians, scientists and teachers would keep these things from you, like how does greenhouse gas affect our planet, so let me tell you a little more.

Stratigraphic Position Timing Issue -Besides Nuclear decay, the main way scientists used to determine "age" was by Stratigraphic Positioning. This is the determination of age by position, depth, and material consistency. MANY TIMES, this is <u>the only method</u> for cross comparison that was thought to be reasonable for confirmation of Radioactive decay. Scientists simply determine the depth of objects, or features near the object, or number of lava flows, or similar geologic characteristics and use the depth as a time gage. This type of comparison may not have a very high level of accuracy, but seeing things in different layers seem to show when something died. If something is lower, it is older and newer is newer. Added to this method is something called the K-T boundary, where iridium chalk was deposited from an ancient meteor that struck the Yucatan around the time the dinosaurs died. Scientists have been using this for a long time when, all of a sudden, there were trees found that were going the wrong way. **Stratigraphic Anomaly**-The next set of pictures shows some of the unfortunate trees that must have died repeatedly to be deposited perpendicular to all of the stratigraphic lines.

Some try to state the trees simply fossilized while standing upright for MILLIONS of years as the ground built up around them. [**20 points on the BUNK meter!**]

Distance to the Sun-If neutrinos from a single solar flare can make things look vastly older, what if the entire Earth was closer to the sun a few hundred thousand years ago? I know that sounds odd, so just keep it in the back of your mind right now. Right now, I'm going to provide you with a more logical way the Pacific Ocean was made at the beginning of the Triassic Period as our planet rotation was not stable. That is where Ice samples come in.

Ice Core Dating Correction

Although the task is tedious, ice can be examined just like tree rings. Each summer ice changes its consistency. H_2O (16) is more concentrated in the summer while H_2O (18) is more concentrated in the winter. This gives us indication to the level of CO_2 which in turn allows us to understand something about the temperature levels. As the yearly cycle has freezing and thawing, ice consistency varies each day, seasonally, and yearly, depending on Earth axis and other critical elements. Anyway, scientists around the world started boring holes in ice. The most coring is done in Greenland and Antarctica. A sample is shown below.

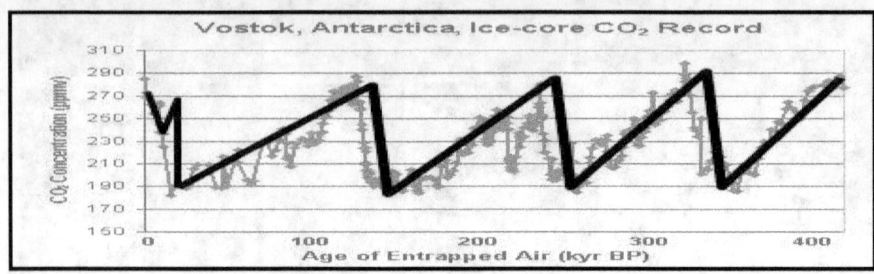

If you look closely you can see that about every 120 years there is a major change in the environment. This was found at both Antarctican and Greenland Ice cores and the dating is by seasonal changes rather than nuclear decay. Bah humb! You say! Well, what if we see confirmation?

Hawaii Hotspot Track Dating Correction

Hawaii is not a tiny group of islands, but instead is an indicator of where the Earth magma has a hotspot. As the crust moves differently than the stuff below, the hotspot relative to the crust moves and each time the hotspot burns through another piece of crust, a volcano erupts which seals off the area after a time and an island is made for a few thousand years. This travelling hotspot known as Hawaii is show next. The descriptions provided shows what was happening along the way. Because the hotspot moves perpendicular to the axis of the Earth, we also know how the earth was spinning as shown by the lines in the first graphic below, but the actual timing is not described here. I placed some general times in the second graphic, but let's see if they make sense.

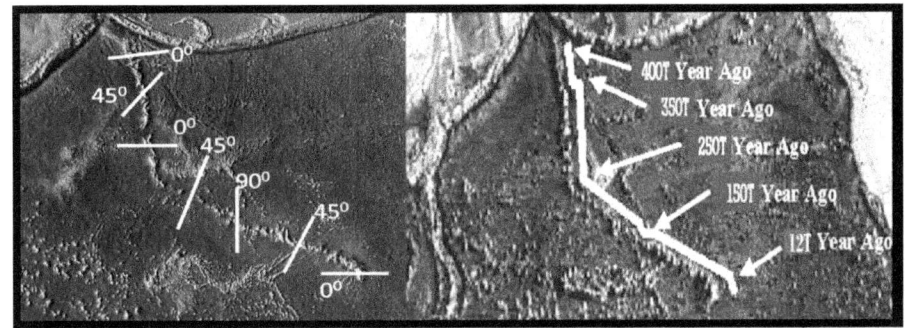

Let's compare the Earth shifts with the Ice core data. Man-oh-man; it seems they match. I think you still believe in nuclear decay so we will look farther.

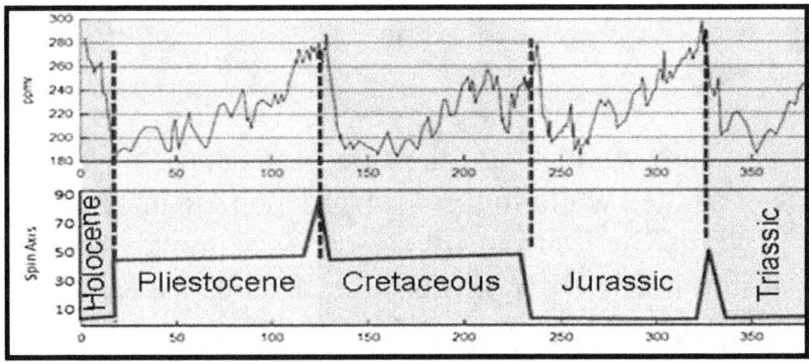

The Atlantic Ocean is getting wider about an inch a year, averaged worldwide. While the building of the great mountains has little to do with the normal tectonic plate "drift" We can pretty accurately measure the widening ocean in various ways including measuring distances between matched magnetic landmarks on either side of a widening gap on the ocean floor. The old theory indicated that 180-million-years-ago the continent Pangea began splitting apart and has been drifting ever since. In so doing, the landmasses of the Western and Eastern hemispheres separated and opened the Atlantic Ocean basin today. Plate tectonics tells us the outer hard crust of Earth consists actually of a dozen or so distinct, hard plates that drift individually on hot, deformable rock. An unequal distribution of heat within Earth moves the plates. The boundary between the plates forming the Atlantic Ocean is smack down the middle along the Mid-Atlantic Ridge, shown as the hashed line in the figure below. The ridge is where we must look to find a widening gap, which accounts for the widening ocean.

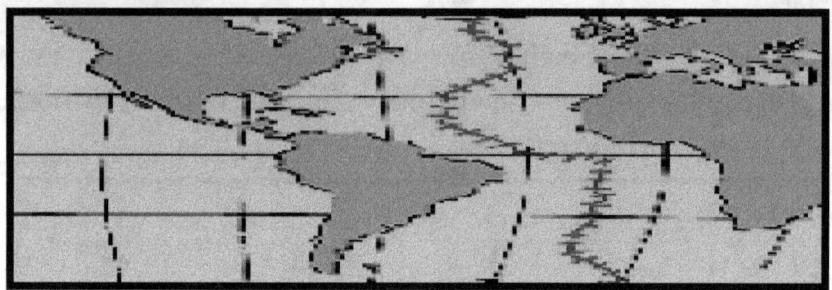

That is where we measure the rate of separation. Where the plates separate, white-hot soft mantle oozes up from great depths within the Earth to fill the gap. The molten rock cools slowly into new slivers of sea floor. This happened over and over again through the eons. That's how the Atlantic Ocean widened-by a spreading sea floor. Iron-rich rock has a peculiar property; heat it above its curie point of 580 degrees Centigrade and it loses its magnetism. When it cools the rock gets re-magnetized in the direction of the existing Earth's magnetic field. So, it's a magnet with the poles aligning with the poles of the Earth at the time of the cooling. The neat thing about this is: the magnetic field of the rock, once cooled, stays frozen in this orientation.

It becomes a record of the Earth's field at the time of its cooling.

The first graph below shows how the magnetic field has changed over time. Certainly, we cannot get an actual time, but a relative timing is very good. What if I told you this matched up exactly with the Ice Core and hotspot data?

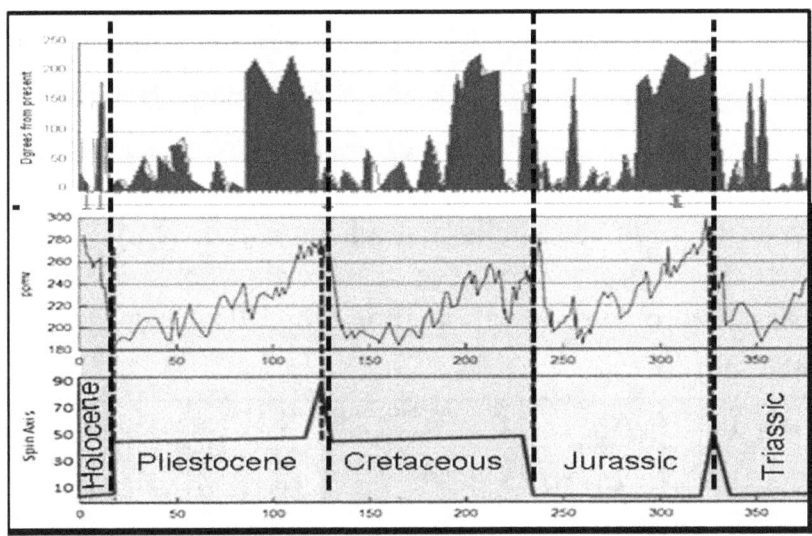

The top graph is a paleomagnetic graph of magnetic field changes over a set time. The cyclic nature generally matches as one would think it should with the timing obtained by the first two ways and it shows that the earth has flipped on its axis many times since Pangea began separating. Besides these three, let's look at marine life.

Marine Isotope Stage [MIS] Dating Correction

Some people may still be reluctant to give up what the schools have been preaching so very long, so I thought I would bring out one last attempt at presenting sanity. Large numbers of scientists around the globe are doing Marine Isotope Stage timing by digging in dirt. It seems looking at the levels of Oxygen-18 shows how hot or cold a point is in time while checking relative Oxygen-18 isotopes in Calcite [which just happens to be the main ingredient in seashells], one can tell just how many of the things were here during each period. Checking around the globe has given us a good map about climate and number of seashells, which correlates to number of animals in general so it is easy to see where extinction periods are. Guess where they line up? Time's up! They are an almost exact match as shown below. MIS levels are shown next above the ice core samples, the hotspot data and the magnetic field shift data Massive drops in O_{18} mean massive drops in sea shells and all other life. Notice there is no extinction period between the Tertiary and Pleistocene Ages marked by Cro-Magnon appearing. Please say you see a comparison.

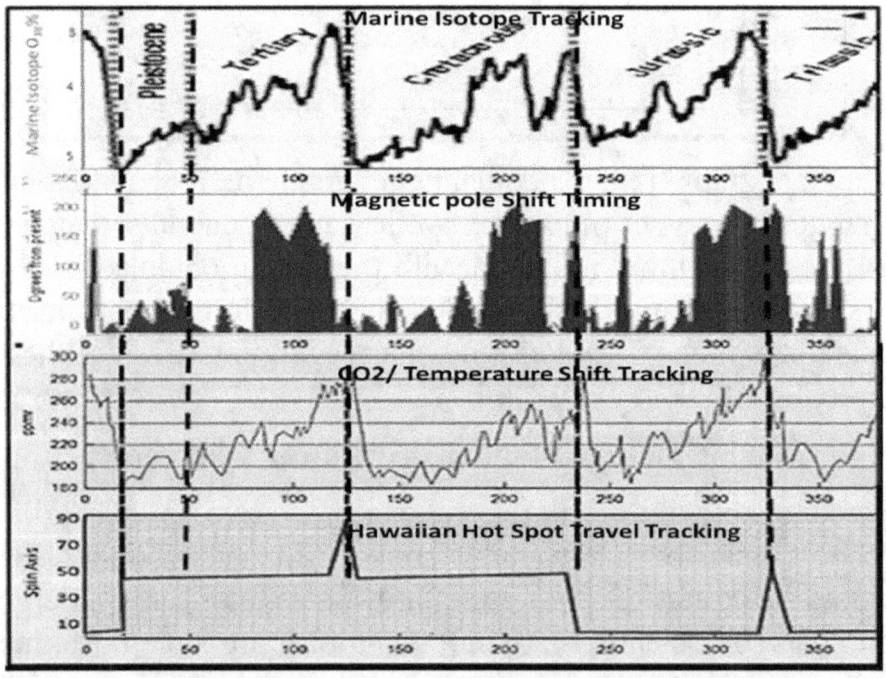

Now for the truth as we can very accurately track seasons etc. in ice core samples so we can get an absolute date to all of this. About every 120-thousand-years there is a massive change in the earth's environment. This is caused by a number of things including a change in the earth's spin axis and other things we seem to have no control over. Typically, we would expect an extinction period whenever one of the abrupt environmental changes happens. I placed the "Common" names for each of the ages with extinctions and while the timings are compressed, the relative characterization of each period is similar to the unreliable nuclear decay method. We find the Pleistocene Extinction occurred 10 thousand years as the earth shifted and shifted back within a very short time. We know the massive herds of millions of Mammoths were instantly shifted to near the North Pole while they were eating flowers and THEY ALL DIED. The next major extinction was the Cretaceous Extinction when most of the dinosaurs died. Rather than being 65 million years ago, it appears it was only about 120 thousand years ago. This helps us understand how and why we are finding dozens of dinosaur-remains that are not fossilized today.

I can't tell you a good reason why teachers are not telling our children about the issues with the old timing and not rushing to this tested, verified, and correlated timing model so we can better understand our world except that books will have to be rewritten and consensus scientists will have to tell the truth for a change.

Earth Stability Fallacy

Besides the compression of timing, if you ask any teacher, and environmentalist and any other consensus scientist, you will hear that the earth is stable and has always been stable as its spin keeps it always going the same way. Unfortunately, not only is it unstable and has had its axis is shifted 170 times during the expansion of the Atlantic Ocean. The reason we use the Atlantic Ocean here is that as the huge hole that was opened up in the Pacific at the beginning of the Permian Age, Pangea split apart is has been slowly trying to cover the hole. After each movement, a hole is opened up in the bottom of the Ocean which fills with magma. As the Magma cools, the iron portions of the magma align to the CURRENT magnetic north. All we need to do is yank out a section and see if the magma lines up in different directions. This is called Paleomagnetic timing. From this timing component we can see that a major change occurred 10 to 12 thousand years ago at the end of the Pleistocene Age as shown below. That being said, the earth did not shift 180-degrees. Instead, it shifted about 30-degrees and, a thousand years or so later, it looks like it shifted back.

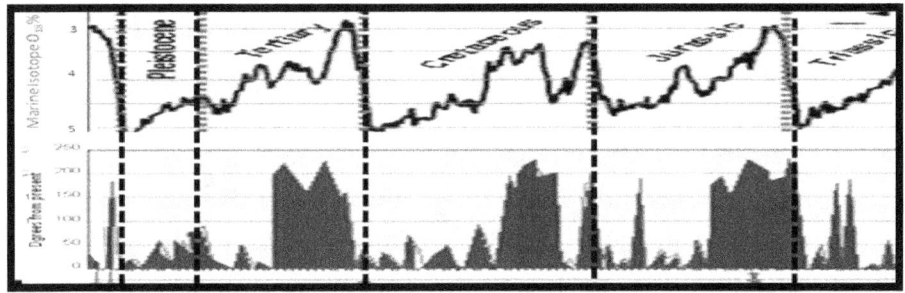

The preceding graph shows how the earth magnetic field provided on the bottom graph according to Paleomagnetic readings compared to the Ice Core dating from Antarctica has similarities that we would expect and the 30-degree-shift lines up with the Pleistocene extinction. If this was the only evidence, we could dismiss it but there are many verifying pieces to consider.

Hawaiian Hot Spot Details- As I discussed earlier, the Hawaiian hotspot not only shows changes consistent with the changes noted in the Ice Core but also the angular changes in the tracks can be reconciled as part of the rea son for extinctions. The turn 10 to 12-thousand-years-ago shows the 30 to 45-degree-shift did occur, however, there seems to be no shift back as the first graph suggests. The images below show the major axis shift and the image to the right shows to possible location of the poles during the Pleistocene that fake scientists are not telling us about.

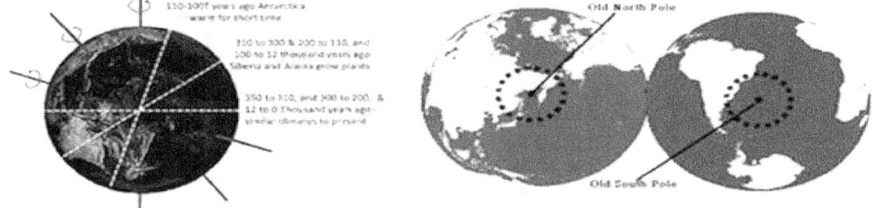

Carolina Bays line of bombardment- The east coast of the United States was pelted with many objects. There are still an estimated **500-thousand** meteorite craters called "Carolina

Bays", which mark this incredible event in history. Some have a diameter of over 14-miles. Just think about how afraid the people of that time were as they essentially saw the sky fall all around them. The picture below left shows the major areas where these objects have been found in the United States. These generally date around the same time and the close collisions tell us the meteor path was along the earth's equatorial path. This shows a shift of 30-degrees from present axis of rotation confirming that when the meteors that caused over 500-thousand craters hit the United States, the earth axis was different. [See the preceding right image]. Evidence in Australia shows a "straight-line" distribution pattern on the other side of the world that is that is consistent with a bombardment along the equatorial boundary. If we consider the impact density line as the "Old" equator, a shift of about 30 degrees in the rotational axis has occurred since the bombardment as shown below. The globe has been separated at this ancient equator. Note that along the equatorial path there was not much land. Also note that eastern Alaska and Siberia are well away from the Arctic Circle, which allowed huge herds of Mammoths to dine on flowers in those areas just before the shift. The shift froze them solid. The next collage shows a tiny area with hundreds of Carolina Bay craters.

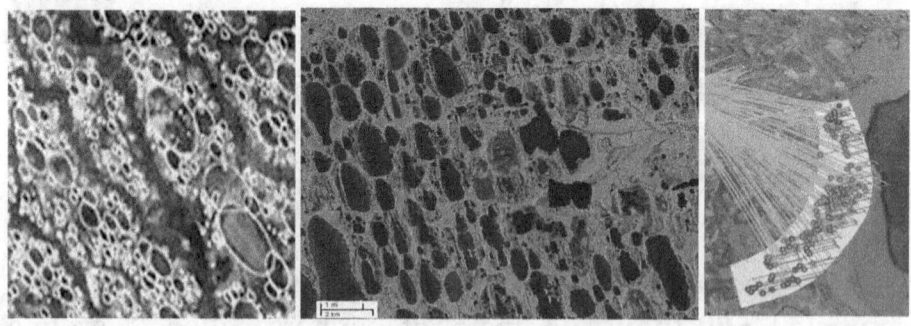

There is something else the Carolina Bay craters tell us. There was a disturbance that could have had enough power to shift the Earth's axis during this time.

Frozen Mammoths- Not only are massive herds of Mammoths found in Siberia where they should not have been, but also some still have flowers in their mouths and there are estimated to be well over a million in the massive herd showing Siberia once was a vast meadow. This means the earth axis was significantly different.

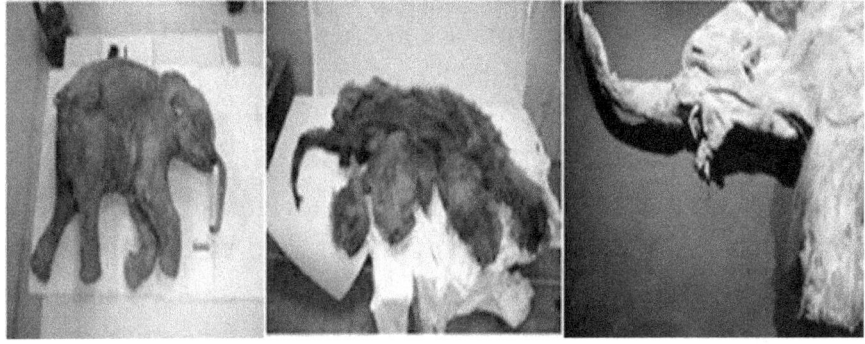

We can almost see the destruction as ½ million meteors hit and turned the forests to fire as noxious plumes rose and the earth shift set up floods, unbelievable tidal-waves that swept the mountain tops, and destruction everywhere. Soon, the entire world would flood as the polar ice caps reformed but many try not to even believe that there was a worldwide flood at all, as the Pleistocene extinction ended. This would give credence to the Biblical history over their own consensus of how the past "should" have been. That brings us to another topic modified by consensus as teachers typically don't even talk about the creation of the Pacific Ocean.

Fake Mystery Ocean

While consensus scientists keep trying to hide the manufacture of the Pacific Ocean, the truth is well known. One day, while Earth was minding its own business, something came along and ripped 1/3 of the entire crust off the planet and it could happen again so we should, at least, know what happened, why it happened, and when it happened so we can be prepared if we have to be. Even after hundreds of thousands of years and after the Atlantic Ocean opened up to try to seal the hole, we can see the huge loss of crustal surface just below the water. [See the left image following.]

 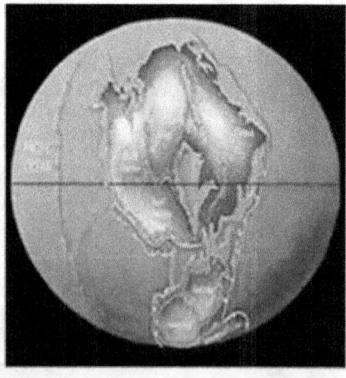

Now there is a huge hole in our planet and no want wants to tell our students that a huge mass was <u>sucked out of the Earth</u> so it is somehow ignored. It doesn't mean it didn't happen, it simply means we lie to our children because we don't like the truth and we think that ignorance is better than being uncomfortable. Below is a topographical cross-section of the earth crust as it is today [across the Tropic of Cancer].

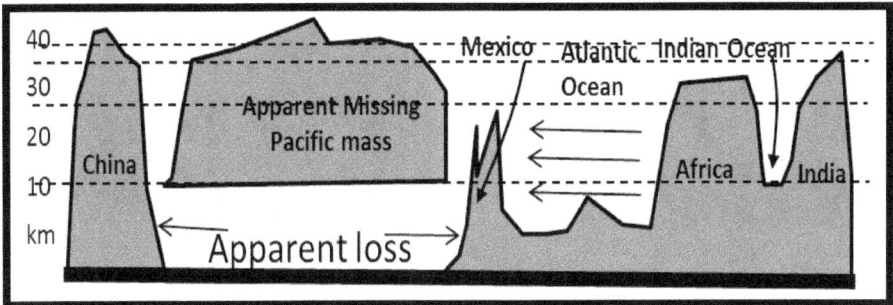

If that was scary enough, the graph below shows that the loss was much worse than it even looks today as well over 1/3 of the Earth's crustal mass was somehow yanked out in space.

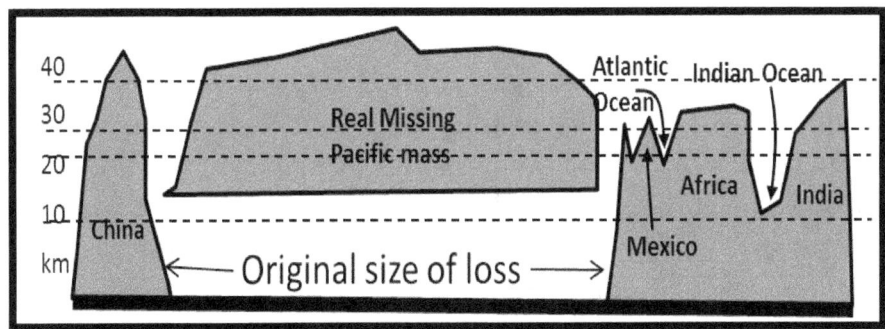

A google Earth view of the Earth today, centered at the center of the hole even after Pangea has been yanked apart to try to heal the massive hole still shows how very horrible this event was. See next left.

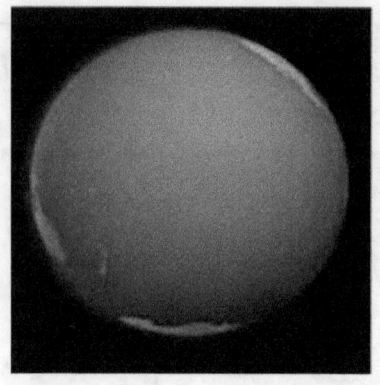 

There is almost no crust below. It's all gone! Teachers and consensus scientists teach a lie by not providing the details. Other works of fiction forced down the throats of children concerning this horror are as following:

Plate Tectonics formed the massive mountain ring around the Pacific Ocean. The naturally formed ocean plates were slammed in every direction at the same time pushing the ocean rim over 5-miles in the air. While there is no trace of the massive volcano or plate pusher in the center of the ocean to do such a thing; don't ask me questions about what pushed the plates. The missing plate pusher is shown in the previous graphic right. Of course, there was nothing to make the plates smash outwards in a circle. Tectonic Forces, typically, work in only one direction, but it didn't fit the "clean version" of our planetary development so this fanciful expanding ocean stupidity is still in the science books today.

Still another crazy theory developed by consensus scientists to hide the truth about our Earth having something such out big chunks is that a massive Island slammed into the land so hard, the 6-mile-high Himalayas were formed and the country of India was made. Strangely India, is made up of almost all magma so how would a magma island move? [See the stupidity below]

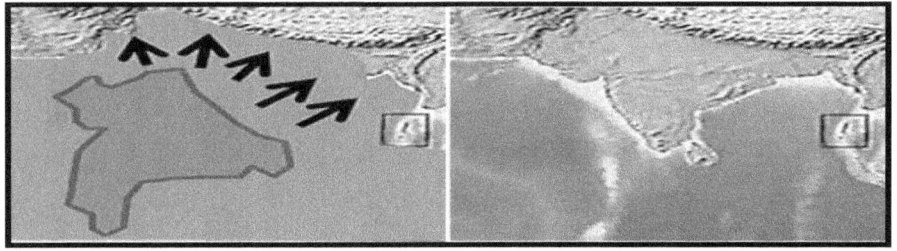

We will get into all of these and many more false teachings built around the desire to keep our children ignorant of the somewhat disturbing fact.

How Much Dirt Did We Lose?

Our earth now has a volume of 288 billion km³. Most of that is called the Mantle with the core also being 38 billion km³ and the crust being about 45 billion km³ [the outside part where people live and plants grow]. Before the event we had an additional 45 billion. Of that part of our planet that is missing, a portion collected to form Lunar, our largest moon; approximately 20 billion km³. The remaining 15 billion went into outer space. No matter how you look at it, that is a lot of rock, dirt, and animals. Even today, the massive hole is so scary just knowing that there is almost no crust below 1/3 of our planet. It's all gone! Teachers teach a lie by not providing the details.

What Other Ramifications?

Lucky for us, the next layer of the Earth, the mantle is much denser and was not affected by the catastrophe. Potentially, one of our massive continental masses on the "Pacific" side of the earth was gone [some call this other massive continent

Prestonia] and only Pangea remained. Water captured under the crust was now free to build deeper oceans.

Just like any other spinning sphere, as our planet got smaller, its rotational speed got slower which would have caused some of our atmosphere to begin leaving Earth just like the rotating Saturn's atmosphere is leaving it today. This slower spin made the Earth gravity less so animals were free to grow larger and larger.

The weather patterns were completely disrupted, so everything about the earth changed. We can believe that not only was rain substantial, but also volcanic action must have been everywhere.

Just like any other mass spinning around an object holding it by gravity, the planet would be driven closer to the gravitational source over some period of time. In this case it was the sun. As the planet came closer, the spin increased which eventually stabilized the gravity, allowed out atmosphere to be regulated and caused a reduction in the size of animals.

A Similar Event

We look at the Planet Mars and the northern half of the planet is almost completely gone. It looks like it was ripped away in a similar fashion to what we see on Earth, but the Planet was not as lucky as the earth as the atmosphere became so limited, it could not even hold surface water. Within a 50-thousand years, Mars would become a dead planet.

The current theory is that Mars and earth came close enough to each other for their gravitational pulls to pull up massive mountain ranges and during a second close encounter thousands of years later, an encounter yanked out the weakened portion of the planetary crust from both planets. The Earth would gain the Moon and the Pacific Ocean, while Mars, got it worse and a massive portion of the planetary crust was ripped away. The planet could no longer hold on to its atmosphere and its water. It slowly became a dead planet. From the change in planetary masses, the orbits of Mars and Earth were separated more than before so that the is a very unlikely scenario that would allow for the two planets to ever come near each other again.

Theories continued to expand and books like *"The Biblical Flood and the Earth Epic"* with editions in 1966, 67, 68, 69, 70, and 1971 continued to tell one of several versions of this horrible event that wasn't based on fantasy like that which is pushed by the Consensus scientists.

Before I leave this section, let me show you the Ice Core CO_2 sample again, but extend it just a little. This is backwards to the others I showed, but don't let that confuse you. Notice that before 450 thousand years ago, the Earth didn't have the massive CO_2 and thermal swings we have seen since. It is as if the Earth was more stable or larger and lost the continent of Prestonia at that time.

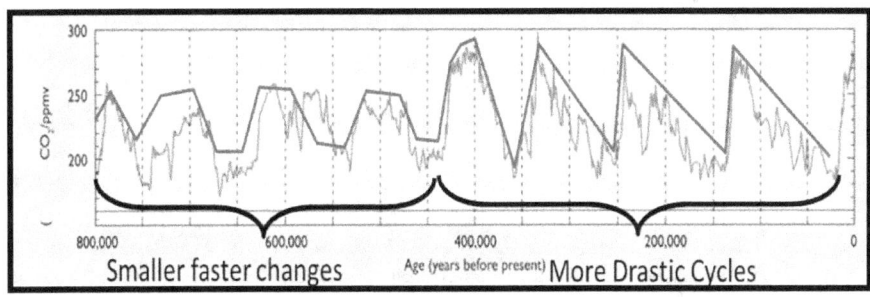

Speaking of the Earth rotating slower so things could be larger, Let's talk about Giants like dinosaurs and humans.

Giant Rejection

While scientists around the globe try to hide the facts, there is absolutely no question that Giant, civilized people that looked just like you and me lived and worked in cities and walked along the same beaches as the dinosaurs during the Cretaceous Period. I'm not talking about the lumbering ape-man like characteristic that also has been pushed either. Hundreds of sites and pieces of physical evidence assure us of this truth. During this time, 16-nuclear plants were established in what would become Gibbon, Africa. The plants are still producing useful fuel today.

Shoeprints- Someone told me that if you find shoeprints there probably was a human foot inside at one time, but consensus scientist can't have that as a truth. Below, we see shoeprints of the civilized people around the world. Sometimes twice the size of a normal person's today, we can imagine the shoe stores were large as well.

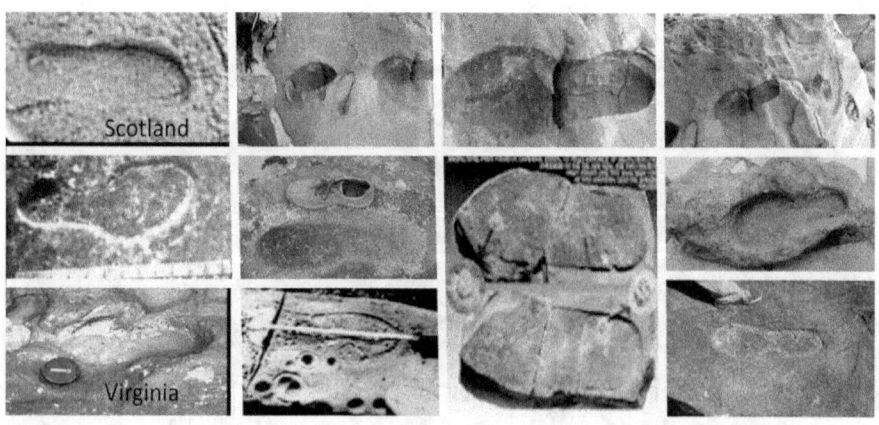

Notice the third shoe on the bottom row. The man stepped on and killed 2 Trilobites that were extinct well before the dinosaurs' Extinction.

Footprints and Teeth-Next is a tiny grouping of the hundreds of fossilized giant footprints and teeth from the time of the Dinosaurs. Many of the footprints are with Dinosaur footprints.

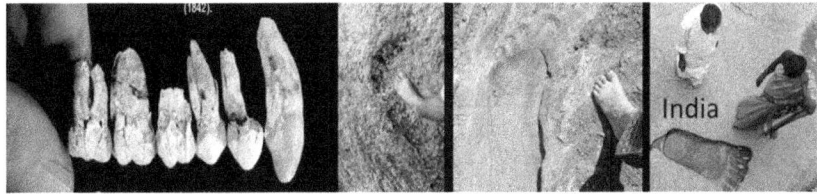

More Footprints- Massive footprints started showing up around the world along areas that used to be beaches.

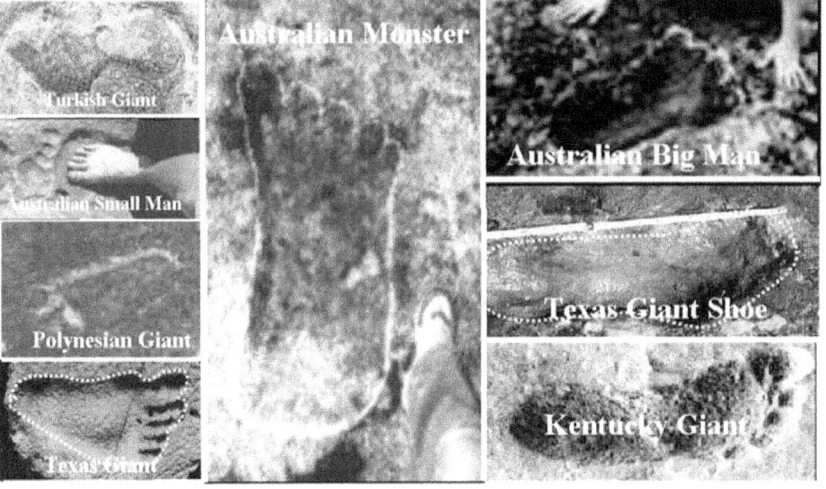

Dinosaurs and People Together-On and on we could go, but the thing that is unusual is that some sites have human footprints and dinosaur prints walked on the same ancient beaches. The graphic below shows some of the trackways.

Petrified-Wilton M. Krogman, the internationally acclaimed bone expert identified a petrified "stone bone" as a human calvarium, a portion of a skull with the eye-sockets broken off. A year later Ed Conrad discovered the large boulder in which was embedded petrified object that bore a distinct resemblance to a huge human cranium.

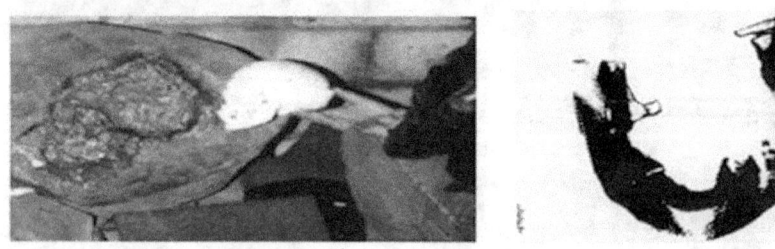

A CAT scan had been done of this particular specimen and revealed intriguing characteristics of a human skull jaw and joints. [See above right]

Turkey- In the late 1950's during road construction in Homs southeast Turkey, many tombs of giants were found that exceeded 12-feet-long. During exhumation, the skeletal remains were examined. The human thigh bones were measured to be 47.2 inches in length. They calculated that the person who owned this Femur, probably, stood at <u>sixteen feet</u>

tall as shown to the left. Images of similar sized giants were shown in the United States, see below middle.

   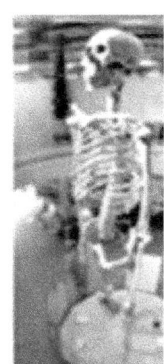

Brazil Giants- Around the world we find these Homo-Gigantus bones. The 2 pictures above right shows a 20-foot specimen.

Manufactured Stones- While I talked about the ancient uranium processing plants that were started by "someone" during the Mesozoic Era; we also, talked about a battery found in a geode showing a level of technology, but something that is still a mystery today is how these ancient people were able to "grow stones" such that there was not space between stones in a wide assortment of placements as wall, floors, and entire buildings found in Peru, Australia, and in North America at West Virginian and Ohio sites. Examples of some of these Grown structures are shown below. It is not known if the giant overlords divulged how to do this or not, but look at the outsides of each of the examples shown. The manufactured stones fit exactly to the adjacent stones and the "skin" of the stone seems to be harder that the insides and grown in layers.

Manufactured Stones

Peruvian Examples   West Virginian Example

Australian Example   Ohioan Example

Sometimes the insides were completely dissolved over many thousands of years as the outsides of these grown blocks were harder than the central core as they pushed against adjacent blocks during the growing phase. [See above]

Manufactured goods- Real researchers have found toys, jewelry, and construction tools. Some of the objects found include the bell, a workman's ladle, and even a hammer with a coalified handle. The last image is some type of valve wheel now integrated with the ceiling of a Coalmine in Russia.

That brings us to many of the buildings from around the very ancient world. The Tower of Babel or Baalbek is a good example of what I'm talking about here. The buildings simply could not have been built without heavy machinery or by giant with capability to levitate to allow for reasonable block placements. The first image in the collage shows that unbelievably massive stones were so easy to place that the builders didn't put them on the bottom, second, or even third row. They placed them on the 4th and 5th level and these things are simply unbelievable at almost 1000 tons. A modern crane tried to move one and simply sank into the ground around the thing. By the way, each of the images below has tiny people in them. These are regular sized men and women.

Then they noticed square holes at the top portion of these gargantuan stones. Square, so holding bars would not roll and positioned so 14-foot humans could carry the blocks to the wall locations. I placed a 14-footer next to a normal guy to show why the holes were so very high.

The giant race of men lived all through the Tertiary and Pleistocene age and into the Holocene, but many texts tell us of the final ending around 1000 BC. That brings us to something the Bible called abominable animals that are also ignored because they don't fit into the consensus. So very much truth is lost while trying to support the unsupportable.

Pleistocene Dinosaur Error

Consensus-scientists agree; all the huge dinosaurs died after the Yucatan meteor hit to end the Cretaceous Period of the Earth and deposited Iridium chalk around the entire Earth. Problem number one is most of the dinosaur carcasses are above the iridium chalk line, so they died BEFORE the chalk marking the end of the Cretaceous Age, but that isn't what this section is about. This section is about an undeniable issue with the above statement as many dinosaur-remains are now being found that <u>are not fossilized</u>. The consensus scientists are going crazy hiding as much as they can. Then there is the verse in Genesis that goes against the consensus of religious leaders that try to believe that the reason most of animals were considered abominations or [unclean] by God was simply because Jews weren't supposed to eat them. The problem is that Noah was told to eat ALL of them, but to keep them separated in his floating ARK.

Genesis 9:3 <u>Every</u> moving thing that liveth <u>shall be meat for you---[including abomination animals]</u>

Some people try to rectify the statement by saying *"God must have made mistakes and didn't like some of the animals that HE had created"*. No matter how crazy that is, at least, they didn't have to read the dozens of texts that indicate why most animals were abominations and why we are finding unfossilized dinosaurs. Many texts tell about how horrible it was to modify Gods animals, but the Giant people were on the earth for tens of thousands of years and would certainly have known how to design DNA modified animals. The religious leaders hide the truth and the fake scientists hid the truth, but now we are finding unfossilized dinosaurs. That means they had to be regenerated during the Pleistocene. The following list shows the abominable animals including all reptiles such as dinosaurs. Whales, Eagles, and Apes were also considered detestable.

Type	Clean or "good" animals	Abominable animals
Bird	Pigeon, Robin, Duck, Dove	Pelican, Ostrich, Eagle, Swan, Owl
Mammal	Horse, Cow, Goat, Sheep, Donkey	Camel, Pig, Monkey, Ape, Porpoise, Whale
Reptile	No clean reptiles	All
Amphibian	No clean amphibians	All
Fish	Only those with Fins and Scales	Squid, Eel, Catfish
Insect	Locust, beetle	All Others That Fly and all that don't

All we can determine since dinosaurs were remade that eagles, Whales, various apes, and other animals were designed by Geneticists. Let's simply review a tiny portion of the written evidence on this subject that is completely ignored by Consensus religious and scientific researchers. While these documents are extremely old and many copies found among the Dead Sea Scrolls, the information can help us tell the truth about unfossilized dinosaurs.

***Book of Giants*-** - *the giants knew the secrets of heaven.* <u>*They made mistakes and they killed many*</u> *animals and people. They selected two hundred donkeys, two hundred asses, two* *hundred rams, two hundred goats, two hundred other beasts of the field. From every animal, and from every type of human was taken its seed for mixed sex. After a time, they defiled the animals and people and begot giants, monsters, and dragons* **[dinosaurs].**

Jasher 4:16-18- *the sons of men in those days took from the cattle of the Earth, the beasts of the field and the fowls of the air, and taught the mixture of animals of one species with the other* **[including dinosaur DNA]**, *for all flesh had corrupted including all animals. ---And after this they sinned against the beasts and birds:* [This sin was the mixing of animal species]

Enoch 4 and 7-*And lawlessness increased on the Earth [War] and all flesh corrupted its way, alike men and cattle and beasts and birds and everything that walks the Earth all corrupted their ways and their orders. And they began to sin with birds and with animals and with reptiles, and with fish.*

Jubilees 5 and 7--*and all flesh corrupted its way, alike men and cattle and beasts and birds and everything that walked the Earth all corrupted their ways and their orders. - Afterwards they sinned against beasts and birds and everything that moves or walks upon the Earth.*

Generations of Adam 6:1-5-*Among the children of Adam, Ammah understood the secrets of creation. She manipulated the very fountain of life until she had created new forms of beings.*

While there are many, many more ignored texts, I think you can get the picture. According to all ancient Middle Eastern histories, most animals were DESGINED by people. The "modified" animals were considered abominations even if they turned out to be Dinosaurs, Dolphins, Eagles, and Apes, but let's talk about the remanufactured dinosaurs that have not fossilized yet.

Today we are finding dozens of UNFOSSILIZED, "REMADE" DINASAURS. The reason we know they were made during the Pleistocene is fossilization would have happened in less than 40 thousand years. The other thing that is telling is that many are radioactive, but that is another story. Here are some more of the tissues found.

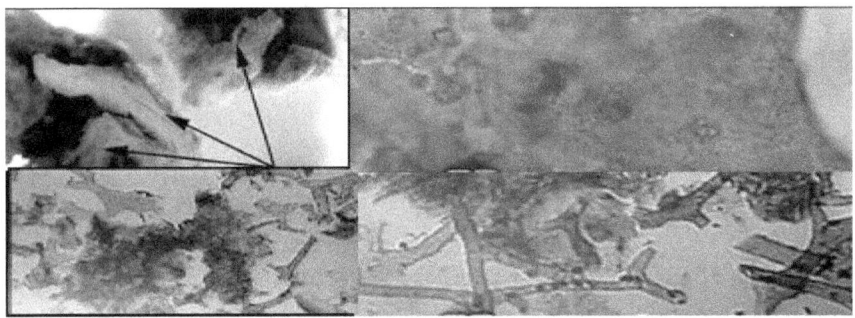

Researchers have discovered flexible and transparent blood vessels, red blood cells, many various proteins including the microtubule building block tubulin, collagen, the cytoskeleton component actin, and hemoglobin, bone maintenance osteocyte cells, and powerful evidence for DNA. Blood vessels from a T-Rex are shown below. In 2012, researchers analyzed multiple dinosaur bone samples from Texas, Alaska, Colorado, and Montana. C-14 dating revealed that they are less than 39,000 years old.

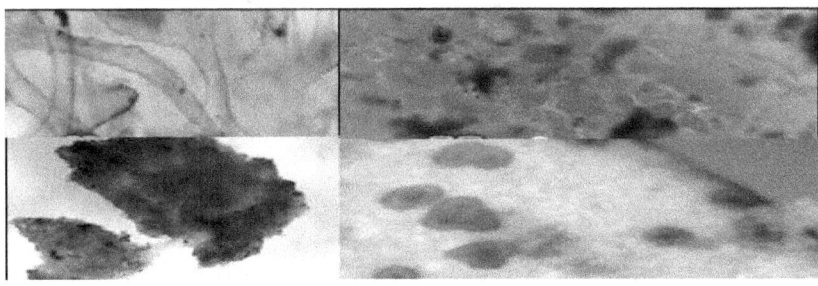

The list of dinosaurs that were, apparently remade during the Pleistocene keeps growing as with all reptiles, they were considered to be unclean and something else was going on. A graphic showing the main ones we know about is shown next.

Every month or so, new finds expand the growing list of remanufactured dinosaurs.

Just think about this. If consensus scientists would just read some of the ancient texts, they would not have to hide the anomalies any longer and there would be a logical reason why people remembered seeing dinosaurs and painted then, carved them, and etched them into wall.

The preceding images from Palestine, Greece, Utah, Colorado, Guatemala, Peru, Mexico, and Cambodia show manmade dinosaurs were one the earth even past the Pleistocene Extinction. An additional group from Peru, England, Spain, Babylon, Israel, Sumeria, China, and Egypt show that dinosaurs were not rare for a while. Also note a couple of things. One is the somewhat smaller Brachiosaurs was the most common type except in Israel and Cambodia where the smaller Stegosaurus and T-Rex roamed. As these images were from 3 to 5 thousand years ago, dinosaurs must have been regenerated even into the Holocene Era.

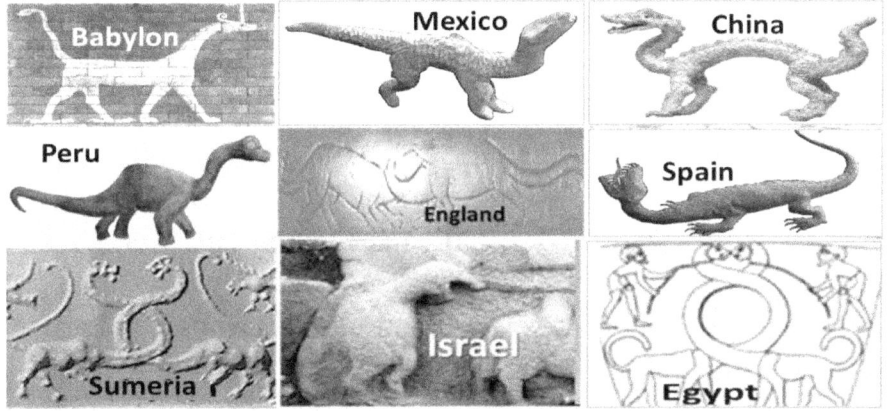

Then there is the book of Daniel in the Bible. This is the last chapter in the King James version of the Bible. In it, King Cyrus thinks his captured dinosaur is a god that can't be killed so Daniel kills the beast. Later versions have taken this part out of Daniel because they simply didn't know about all the other evidence, but we are talking about dinosaurs alive in 600BC.

***Daniel [Bel and the Dragon] 14:23-28-** And in that same place <u>there was a great dragon, which they of Babylon worshipped</u>. And the king said unto Daniel, Wilt thou also say that this is of brass? lo, he liveth, he eateth and drinketh; thou canst not say that he is no living god: therefore worship him.*

Then said Daniel unto the king, Give me leave, O king, and I shall slay this dragon without sword or staff. The king said, I give thee leave. Then Daniel took pitch, and fat, and hair, and did seethe them together, and made lumps thereof: this he put in the dragon's mouth, and so the dragon burst in sunder: and Daniel said, Lo, these are the [fake] gods ye worship. When they of Babylon heard that, they took great indignation, and conspired against the king, saying, The king is become a Jew, and he hath destroyed Bel, he hath slain the dragon, and put the priests to death.

The priests got their way and threw the dinosaur killed in a lions' den, but he was not eaten. Many in the ancient world saw and even killed dinosaurs--------. There is a reason dinosaur remains are not fossilized. --------Some were still here even less than 3 thousand years ago. There is a reason why whales, and apes, and dolphins are considered abominations. ------They were modified from some other animal. The reason many T-Rex remains are being found so radioactive that they must be painted with special Lead Paint may have something to do with these remade dinosaurs and something called the Young Dryas.

Hidden Dryas

In this section I'm going to tell you about something scientists call the Young Dryas along with the destruction of Venus, but scientists don't seem to want to tell you what must have happened as it involves a nuclear world war, the Earth axis shifting, space travel, and Venus. This was a strange time just before the end of the Pleistocene Extinction and worldwide flood. I already mentioned the ½ million craters called the Carolina Bays and, in this section, we will learn more truths about them as they occurred during the Young Dryas the fake scientists are trying their best to hide. These scientists want to hide the known events and conditions and simply say <u>Dryas are a mystery because they mess up their predefined world, because it disproved evolution, eliminates the greenhouse effect fear, destroys their image of how man developed, shows how nuclear events mutated humans so many years ago, and just about everything else they have been cramming down our throats.</u> The graph following shows the dramatic decrease in temperature between about 11 thousand and 10 thousand years ago. The dryas ended with the Pleistocene extinction, but how did it begin?

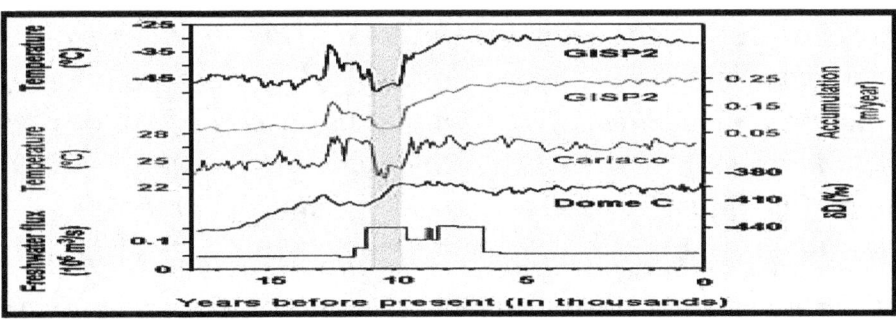

The top two ice core samples are from Greenland and the last 2 are from Antarctica to show this was a worldwide event. While we don't know for sure what exactly happened, here are the things we know.

Young Dryas Radioactivity*- We will see there was a sharp short-term rise in radioactivity about 11 thousand years ago. Uranium concentrations in coral jump by almost 300% during this time and we also are finding unfossilized dinosaur remains that are highly radioactive and knowledge of 16 nuclear processing areas in Africa predate the end of the Pleistocene. The following chart shows the many indicators of Nuclear events during this time including nanodiamonds, and carbon spheroids.*

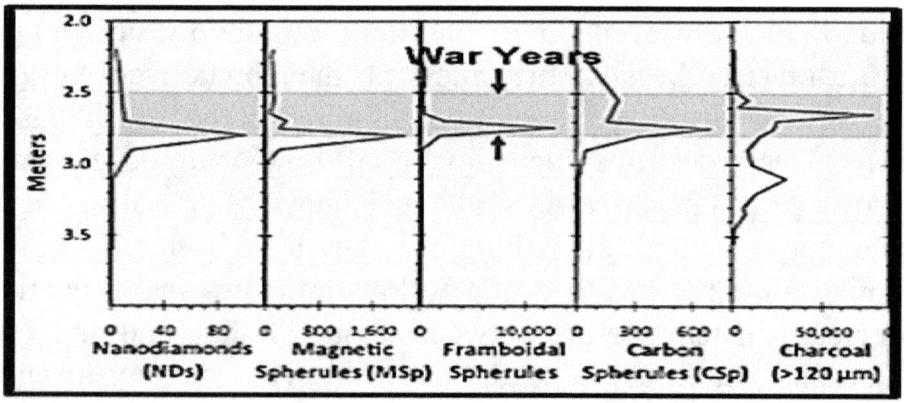

History*-There are many depictions of a massive worldwide war just before the world wide flood. This includes stories from our Bible indicating as many as 1/3 of the population of the world was killed. According to those records, the wars began around 15 thousand years ago. Zoroastrian, Greek, Sumerian, Indian, Chinese, and other histories confirm it.*

Plato*-Tells us of a group of Islands that sank, before the end of the Pleistocene. From ancient Egyptian records, he timed* the submersion to be about 12 thousand years ago. Egyptian history called the Emerald Tablets, confirm the sinking.

Caroline Bay Craters- ½ Million meteors came from somewhere around <u>11 thousand years ago</u> and it may have been the catalyst that would shift the Earth's axis a thousand years later. The meteor bombardment must have taken at least an entire day and maybe more. Also, the disruption that caused the meteors was fairly close [possibly only 30 million miles away, near Venus] as the meteor-scatter is confined to the ancient equatorial area.

Temperature-There was a massive drop in Earth temperature around the world about 11 thousand years ago as if some catastrophic event like a multiple-million meteor bombardment occurred.

Venus Rotation- In fairly recent times, Venus rotation shifted 90 degrees from all the other planets. As it is perpendicular to its revolution, it is known that something huge hit it and shifted it rotation.

Venus Recent Catastrophe- Massive rivers look just like they did before all the water was sucked by the 800-degree temperature that occurred after the planet shifted. Volcano lava flows were poured out and solidified as if they happened only a short time ago.

Venusian Atmosphere- still contained Argon, methane, and Ozone showing the event happened very recently and that some type of life is still on the planet producing Ozone and methane.

Equatorial Craters- We are finding that most of the major craters on the planet fall along the equatorial region as if whatever caused them at one time was extremely close; possibly a moon.

Planetary Colonization -Physical evidence and written indications that the planet Rahab [Venus] was colonized as can be found in many ancient documents.

Venus War- We are now finding many signs of war on the Planet Venus. Many are staying quiet about it, but these are being timed as "recent" by astronomers and researchers.

Biblical Testimony- Clear indication of the destruction of the vain [rahab] planet can be found in our Bible and other histories.

Pleistocene Weaponry-Descriptions of the use of mighty weapons in a Pleistocene War are found in a number of Bible related works which include flying vehicles capable of space flight and massive weapons of war during the Pleistocene.

Pleistocene DNA Mutation- We are told by Haplotype scientists that almost ½ of all major DNA mutation occurred 10 to 12 thousand years ago and most of the other mutations occurred around the time of the Bharata War. The massive mutations are a major sign of Nuclear War around 11 thousand years ago.

Plasma Ribbon- The remains of a plasma string still extend from Venus to within a short distance of our planet, according to the Soho Satellite data. Plasma makes a fluorescent light, but when it is massive, it is another thing. When this plasma string once connected our planets momentarily, a massive lightning-bolt unlike anything seen recently would have discharged our 2 planets' electrical charge and we are told the plasma string would have been visible for many years as the energy dissipated after *separation. Dozens of eyewitness accounts of Venus having a red flowing tail shows this event occurred in the fairly near term [possibly 11 thousand years ago].*

Plausible Explanation

These things seem to tell us there were civilized people on the earth during the Pleistocene who experimented with DNA remaking dinosaurs that had once been extinct from the ending of the Cretaceous. Desire for power abounded and there was a war before the worldwide flood and many were killed. At some time, nuclear weapons were used. During this time the planet Venus was livable and colonized, but some terrible even destroyed the planet and sent massive amounts of meteoric debris to the Earth. This meteor shower upset the weather patterned and Islands became submerged. I know some of this is not being taught in our schools and I don't have a problem with that so much, but what I have an issue with is that the things we know are not being taught.

Kingdom after kingdom became accustom to war. Some flying ships were sent to remote sites on the Moon and Venus according to many sources. Venus was not like it is today. It had been a lush planet with green fields, huge rivers, massive oceans and air. According to the Biblical account, the Planet Venus was called Rahab or "Vain Place". It also states that one of the warring generals of this time, originally named Gadrael, was massing a large group to assault "heaven". The book of Ancient Jewish history, *Jasher,* indicates that *"1/3 of the population of the entire world was killed in the World War events." To give you perspective, World War I and II combined killed about 1.5% of the population.*

The ancient historical works tell us that the planet Venus was much larger in the sky than it is today and both planets affected one another somewhat like the earlier events of Mars, but not to that extreme. Venus would have been the best colonization port as it most likely had sufficient atmosphere, oxygen and water to support a large civilization and it was only 5 million miles away. The remains of huge river systems, rolling hills and valleys can still be seen today, but there are no more people and there is no more breathable atmosphere, in fact, it is almost all CO_2. There also has been a huge temperature jump to over 800 degrees, the air pressure has jumped to 90 times that of the earth's, and thick clouds of Sulfuric Acid now cover the surface. You can, certainly, believe that anything that had been alive on the planet was destroyed in the transition. Another thing that is noticeable, is that the rotational speed slowed to almost nothing. The logarithmic chart below shows the rotational ratios to planet size of the other planets. Notice that Venus is way out of place. It has almost no rotation. In fact, Venus is currently rotating about 1/100th as fast as its sister planet Earth.

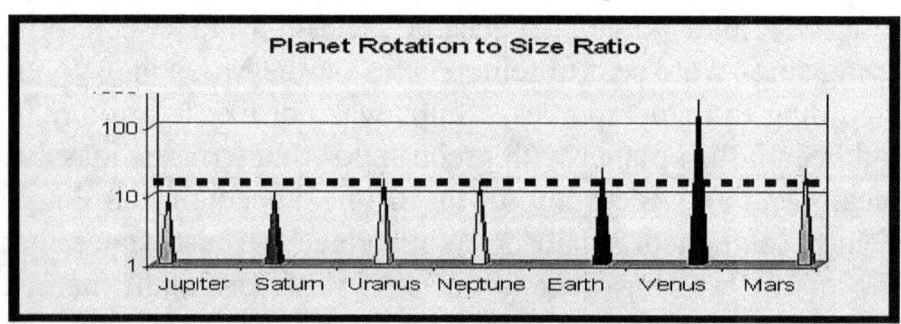

Something catastrophic happened on Venus but we will only investigate it superficially, as it affected those on the Earth during the World War before the Pleistocene Extinction. The image following graphic left shows what Venus looks like today with most of the massive craters along its equator?

 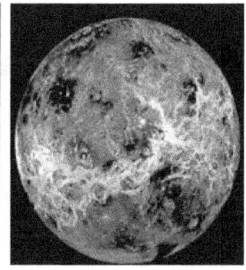 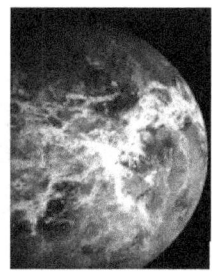

It was as if its moon soddenly exploded and scattered debris along its path on the surface. This would have enormous consequences. If we look closely, we see the planet almost split in half [second image shows white where a massive gash is visible], and we may even know where the impact as X marks the spot [last image above]. This drove the rotation to an almost stopped condition, causing the atmosphere to build until the pressure level hit unbelievable levels, structures melted, and rivers died up immediately leaving well defined images of their once great streams. The end of a livable Venus may have looked something like this. Scientists have known these things for years.

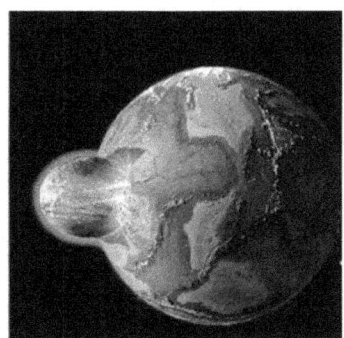 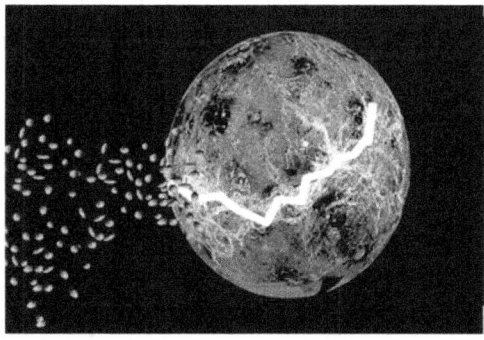

Why- We may have some information about what could have caused the moon to explode. It got too close to the Earth and when its moon got between the planets, it exploded. The impact shifted Venus' axis and destroyed everything. The remains of a plasma that could cause this type of discharge is still visible. The Soho Satellite registered a substantial indication of a plasma "electrical discharge string" emanating from Venus and going towards our planet. The image below shows the satellite position when it captured this unbelievable 30-million-mile long electrical plasma emanation. *Scientists have known this for years. They also know about dozens of sightings of Venus showing a visible tail for a number of years, but they don't want you to know.*

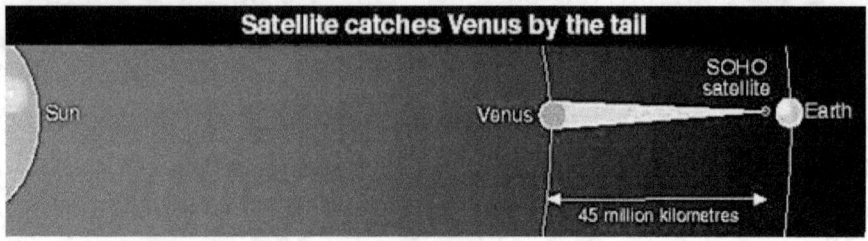

Venus Today-By far the most noticeable thing on the planet is lava. The collage shows just a few of the volcanos that pot mark the surface. While they appear to be spewing lava out for hundreds of miles making it hard to walk even with shoes on, this odd thing is the <u>eruptions happened all at one time</u> and the outcome is frozen in time. The detail left by the eruptions show that the incident did not happen long ago. Scientists tell us these are <u>new eruptions</u>.

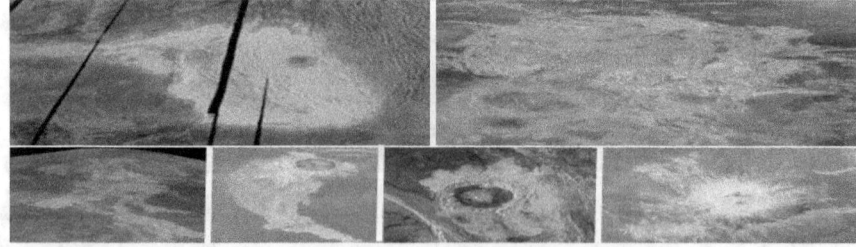

Besides the Sulfur peppered Carbon Dioxide air, we find what looks like snowcapped mountains. What appear to be snow-capped mountains is actually metal topped mountains where the metals have melted to build a highly reflective surface feature. One of the many beautiful mountains is shown next.

The next thing we find is what appears to be remains of buildings beside massive waterway now completely dry. The next image shows the anomaly. I enhanced the areas of interest to show how it used to look below the original.

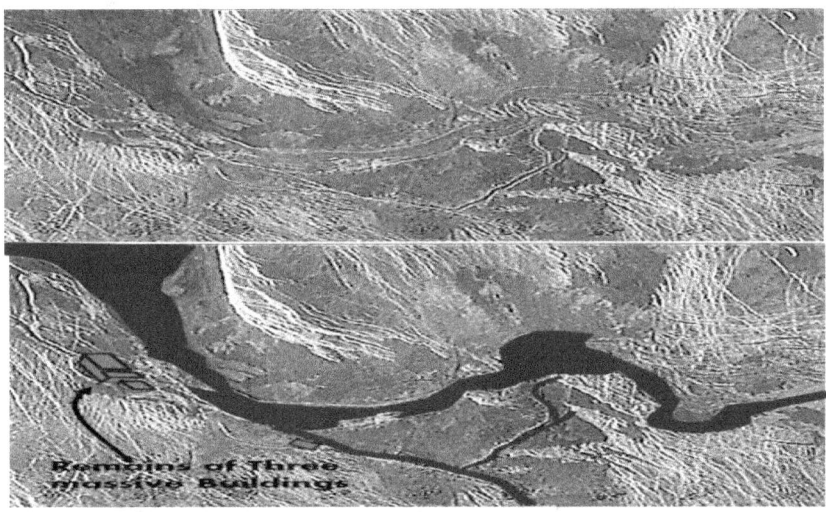

Rivers-The whole countryside has the remains of massive waterways. Look at the following river delta and you can get a sense of the beauty Venus must have had at one time. The huge quantity of tributaries in the image right means massive amounts of water flowing from the mountains as rain pelted the countryside.

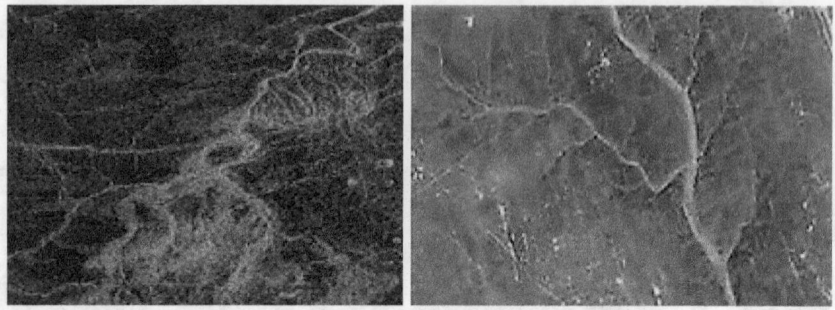

Roads-Over the tops of all types of hills and valleys, Venusian roads seems to go on forever which shows a reasonable probability that someone used those roads.

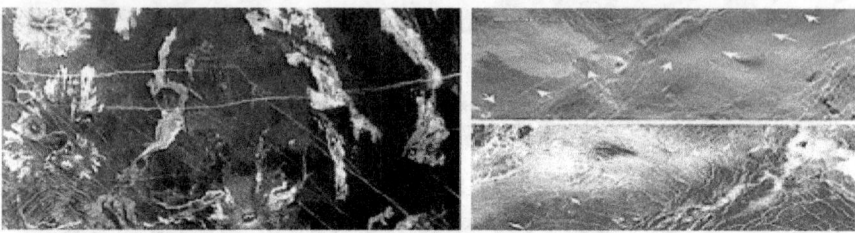

Remains of Cities- In one of the protected valleys we even find what might be the remains of a city before being melted by intense heat and pressure. Remains of possible large buildings also can be found.

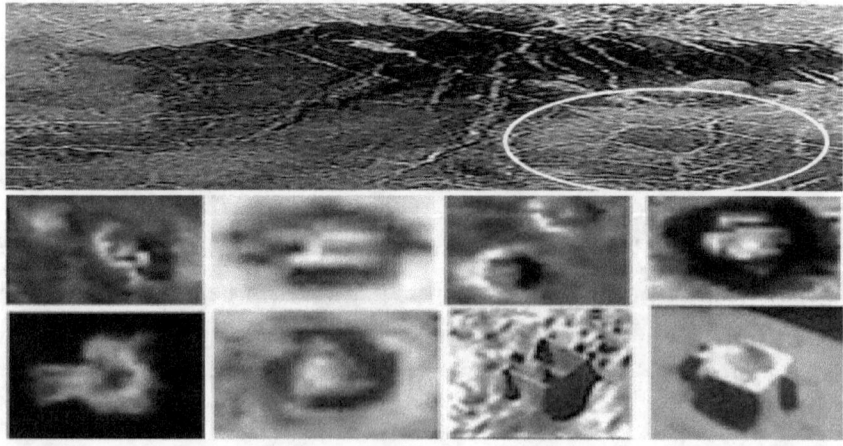

Space War- Back on earth, the wars are really bad. The book Jasher indicated 1/3 of the inhabitants would die. They were getting worse and worse and there may have been continuous fighting for almost a thousand years. By now many people had moved underground. Soon, attacks on the planets and on the moon were in fashion. In India there is written evidence that battles were fought on the moon and in China descriptions of methods to transfer a detachment of men onto another planet has been found. The most likely planet would have been Venus. These descriptions and those below may paint a strong picture that the fighting was not confined to the Earth.

***Indian Version**-Maharishi Bharadvaya- There were gigantic battles in heaven.*

Babylonian Version-In the "Epic of Etana" we read, *"Etana looked down and saw the Earth had become like a hill and the sea a well and so they flew for an hour and Etana looked down and the Earth was like a grinding stone and the sea like a pot. After the third hour the Earth was only a speck of dust and the sea no longer seen"* [The ship, of course, was going into outer space.]

***Chinese Version- Lhasa Sanskrit detail**—the builders of the "Astras" could have sent a detachment of men to any planet. --*

Greek Description-From Greek legends talking about battles between the gods we are told the following: *"Hot vapor lapped the titans, flames unspeakable rose bright to the upper air [outer space], lightning blinded their eyes."* Apparently, lightning weapons or weapons that affected plasmas were used in outer space.

Venus battles- By accounts in the Bible and other places, there evidently was a very active group of warriors lived on Venus so we can expect it had its share of attacks. From the looks of things, the destruction of Venus was not immediate. Instead, bombs blasted the surface. The following image is a good place to start.

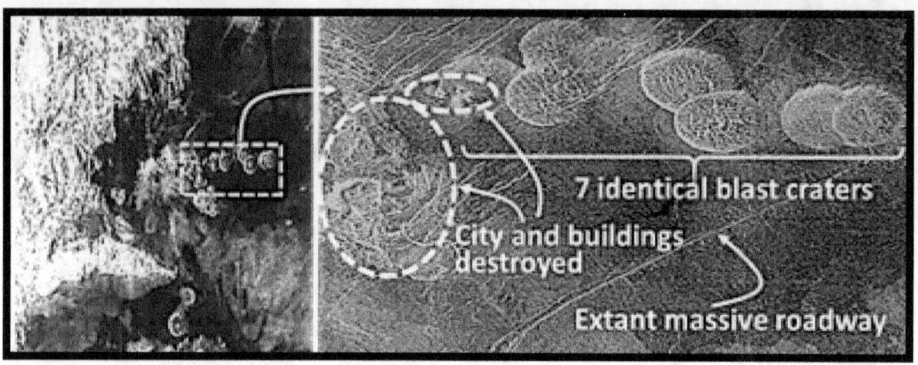

The image above is approximately 350 miles across shows smooth volcanic planes that border the eastern edge of the Alpha Regio Mountains. Within these dark plains some very strange pancake domes that are about 20 miles and the crater edges are about ½ mile high. That's not the unusual thing. While much of the planet's artifacts have melted away in the intense heat that is consuming the planet, some evidence remains to this day. Here is a blow up of the preceding image. Some try to ignore these features or come up with crazy explanations that don't make them say the war word, but hiding information does no one any good.

This row of IDENTICAL craters is not indicative of a meteor shower that would cause a random layout of variably sized blasts as the meteor exploded high in the atmosphere. These strikes are directed in a line on Venus. Here are seven identically sized blast areas in line. Someone was apparently trying to hit something during a strafing run of some kind. One thing that should be noted is that each of the blast areas is exactly the same size so the blasts could not have been random pieces of meteor unless each piece came from the same source, all happened at the same time, and all pieces were the exactly the same size and density. This is indicative of bomb blasts not meteors or volcanoes. Please notice, also that the blasts are along a very distinct roadway and if you look closely it appears that they hit some objects that were rectangular as the blow-up to the right shows where 2 of the blasts struck.

Biblical Description- *Psalm 89:5-11 LORD -- You have broken Rahab [the vain planet Venus] in pieces, as one who is slain; you have scattered your enemies [*Satan's army colony]. *The heavens are now yours.* Rahab was in the heavens and had some of Satan's Army on it and now no one can use the planet.

Biblical Description- *Job 26:12- Satan and his followers rebelled again [during the Pleistocene]. God destroyed the army of Satan's dwelling places. He smashed Rahab. It was reduced to **stones of fire** Stones* of fire is code for a meteor storm that blanked the US East Coast.

Biblical Description- *Enoch 85:1-4 -A single star fell from heaven* [destruction of Venus]—*then I saw many stars* [meteors] *which descended and projected themselves from heaven to where the first star was.*

Biblical Description- *Isaiah 14:12-14* How art <u>thou fallen from heaven, O 'Morning Star" [Venus],</u> *son of the morning! how art thou cut down to the ground, which didst weaken the nations!* **51:9-10** *Awake, O LORD! Awake as in the ancient days. Are you not the <u>arm that cut Rahab apart</u>?*

Persian Description-Nestorian Library-Rahab Psalm--*My God, <u>Thou didst crush Rahab like a carcass; thou didst scatter</u> [Meteors] <u>Thine enemies with Thy mighty arm.</u>*

Tibetan Description- "***Book of Dzyan***". *The <u>Guardian of Venus</u>, with a mighty roar of swift decent from incalculable height was <u>surrounded by blazing masses of fire which filled the sky with shooting tongues of flame.</u> The vessels of the lords of <u>the flame flashed through the aerial spaces. Stars showered down</u> on the black-faced while they slept. A large heavenly stone crashed into the Earth and caused a long time of darkness and cold. Thus did the <u>large, lumbering creatures come to the end of their season. The first great waters came and swallowed the 7 great islands.</u> The <u>holy were the only saved</u>. The unholy and most huge animals were destroyed. Few men remained. <u>The guardian of Venus came down to Earth with his disciples in huge ships.</u> The king of the Dazzling face, sent <u>his air-vehicles</u>, with pious men inside, to all the other chiefs saying, "Cross the land while dry. The lord of the storm is approaching." Let the dazzling faces take the flying ships away from the lords of the dark skinned. The waters had already moved, but the nations had now crossed the dry land and were led to the lands of Fire and metal. <u>The waters covered the whole world</u>.* [This is a great description. There were many explosions on Venus. The explosions fostered many meteorites, which filled the earth's sky. Meteors from Venus hit the earth and it was followed by a shift in our axis. This is saying the remade dinosaurs died--- we can be pretty sure none were in the Ark of Noah. Some were warned of the great catastrophe. The giants were destroyed and some survivors from Venus came back just before Venus turned into an inferno. Some of the ancient people escaped the doom of the worldwide flood by using the flying ships that I mentioned before.]

Brazilian Description- *In the west, where now is only water there was a large island. A second gigantic mass of land was in the northern part of the ocean as well.* <u>*Both lands were buried under an enormous tidal wave during the first Great Catastrophe.*</u> *It occurred* <u>*at the end of the war between the two divine races*</u>*. The war between the two divine races did* <u>*not only lay waste to the earth,*</u> *but* <u>*also to the worlds of Mars and Venus*</u>*.*

Chinese Description-Sanskrit Documents from Lhasa -- Methodology of *how to* <u>*send a detachment of men onto any planet*</u> was described.

Greek Description-From Greek legends talking about battles between the gods we are told the following: "*Hot vapor lapped the titans, flames unspeakable rose bright to the upper air [outer space], lightning blinded their eyes.*" [Apparently lightning weapons or weapons that affected plasmas were used in outer space.]

Ebionite Christian Descriptions- ***Origins of the World-*** *Before the end of the age [Pleistocene Age], the* <u>*world shook with great thundering*</u>*. Then the kings* <u>*waged war*</u> *against one another. The* <u>*star of the sky*</u> *[Venus] canceled its circuit. and* <u>*from the heavens fell upon the next forces and were consumed*</u>

by fire. Everything seemed good but <u>only to be replaced by</u> <u>the **spawn of a lower star**</u> [many meteors from Rahab]. *Then, over the world <u>broke the great waters, drowning and sinking,</u> <u>changing the Earth's balance.</u>*

Witnessing the Destruction of Venus and Plasma Tail

New Zealand Tradition-The Maori tradition indicates that *long ago a glowing object appeared in the sky, then shattered, showering the Earth with devastating fragments.* [They saw the destruction of Venus]

Chinese Memory -According to the Chinese historian, "Chuangtsu", "*The earth was struck during the reign of Emperor Yao, an intense heat lit up the earth, frightful hurricanes destroyed cities, the sea rose and boiled, submerging fields*"

Chinese Venusian Meteors- The Chinese writers said the same thing, "*There was a time when a planet [Venus] approached close to the Earth, causing great showers of stones.*"

Indian Venusian Plasma-The Indian writers also informed us of this terrible calamity. "*Her [Venus's] anger grew so terrible **her hair is wild**,* [Here we read about the comet-like tail and so many meteors that the sky is darkened.]

Central American Venusian Tail- The Aztecs called **Venus** "*the Star that smoked*" and said that *it once passed by the world blazing and killing many people*

Travel in Space

Indian- "Ramayana"- The "Ramayana" said the following, "It was a self-sustaining flying city that **traveled in outer**

space"-- "One of these cities was named Hiranyapura (city of gold)"

Indian- "Amsu Bodhini"- Information about planets and machines to carry people to other planets was found**.**

Biblical Description of Pleistocene Wars

Jubilees 5 and 7 - *Lawlessness increased on the earth and all flesh corrupted its way and they began to slay each other till they all fell by the sword and were destroyed from the earth. --And God destroyed all from their places, and there was not left one of them. ---and the Giants slew the Ancient humans, and the Ancient humans slew the Eljo [other half breeds], and the Eljo mankind, and one man another.*

Enoch 10:13-14-*Destroy the children of fornication, the offspring of the ancient people, from among men; bring them forth, and excite them one against another. Let them perish by mutual slaughter.*

Generation of Adam 11:3- *"Leboa, Daughter of Tamar, daughter of Adam, devised a "Sword of Light" which penetrated the wall of defense around the city of Haner and began to drain the power from the wall."* I know this sounds like Star Wars, but it is talking about war during the Pleistocene.

Hesiod [Greek]-- *All the gods were divided in strife, even to mingle storm and tempest and already hastening to make an utter end of the race of mortal men, declaring that he would destroy the lives of the demi-gods, that the children of the gods should not mate with wretched mortals.*

Hopefully, you are at least intrigued by the data that teachers have not provided to students so that they could make some logical determination of the events that molded Venus, sunk the island of Atlantis, elevated the radiation level of the earth, made the temperature drop suddenly, caused massive amounts of mutations in humans during a very short period of time, cause many dinosaur bones to be radioactive, caused the plasma strings being seen by Soho, caused 500 thousands craters along the East coast of the United States, and was described over and over again by people around the world. Please do not just believe CONSENSUS. The scientist using this horrible tactic have an agenda and it is not to inform people about the truth.

Why Lie?

Fake scientists have a lot to lose as they have been ramming down our throats their consensus description that elevated their bogus theories. If their theories are wrong, they might not be as mighty as they think. All mankind gets the brunt of this insanity and because we let them do it, they push harder and harder to make themselves the arbiter of all things scientific.

Fake Historians

This not only happens in strict sciences, but also in all out lives. Let's take, for instance, the Crusades. What we are told is the Christians bellowed out of control yelling death to the peaceful Muslims and killing everything in sight showing Christianity is no better than Secularism and man does not need God. The hidden truth is much more complicated and it does not suggest the removal of all things God.

Mistaken Crusades

Before I end this book, I want to address one other consensus issue that should be disturbing all of us, but instead, it is being hidden deeper and deeper each day. I'm talking about whitewashing the viciousness of the pagan religious cult to Allah' the moon god so worshipping the Creator seems inappropriate or wrong. What we find, instead is the Muslims invade Europe and almost take control of all the Christian lands over four- hundred-year period before a desperate group try their best to retake the land. Of course, like all things, it isn't that simple, but worshipping the Creator had very little to do with any of it.

This story is about a cult that requires its believers to try to get to Mecca at one time in their lives to kiss a meteorite that had been worshipped for 700 years before the Muslim religion was established. The stone is called is called "Yamin Allah" (the right hand of Allah'). By the way, the name Allah' is a variant of the Arabian god "Allat" not Yahweh. Originally their sun god had three daughters and Allot was one of them, but by the time of the Greeks, according to Herodotus, the ancient Arabians believed in only two gods: *They believe in no other gods except Dionysus and the Heavenly Aphrodite; and they say that they wear their hair as Dionysus does his, cutting it round the head and shaving the temples. They call Dionysus, Orotalt; and Aphrodite, Allat.* Slowly Allot became male and the name was slightly changed to Allah.

Here is the weird part. Aphrodite had an anthromorphic form of a meteorite called the Aphrodite Black Stone [below left]. Roman coins showed the sign for Aphrodite on their coins as a crescent moon or crescent moon with a star above as shown next. Next to those are the image of Ishtar with a crescent moon above her head followed by Isis with her crescent moon at the base of the sun, and the Greek Artemis with a Crescent moon above her head. The 4th one is Luna Roman Goddess of the moon and the last is Selene Goddess of the moon in Greece. It can be easily seen that these are all the same person and the Moon.

Muslim Cultists are supposed to come to Mecca sometimes in their lives to kiss the meteorite stone of Allah'. This is a very big deal as the building holding the Meteorite is draped with massive tapestry [see 1st below], then hundreds of thousands of followers circle the building three times before finally kissing the rock as shown. [The meteorite in inside the white thing.]

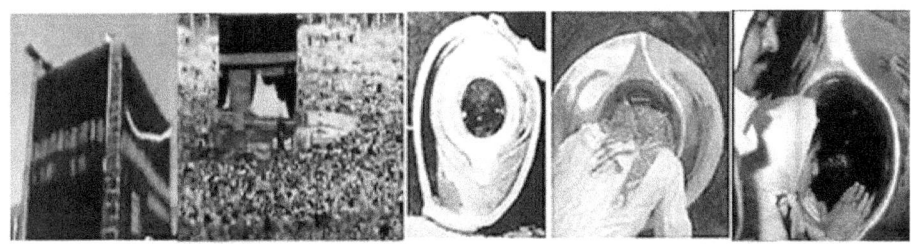

To identify Allah with the moon, thousands of shrines and flags of the Muslim countries adorn the crescent moon. The images below show Assyrian relief depicting the moon god of Babylon, the Ancient Mesopotamian Crescent and Star of the moon god called "Sin" and Muslim mosque images.

To this day, Allah's "Moon god" origins are visible on the flags of every Muslim country. The next image shows just eleven, but all worship the moon god.

I listen to the television today and the news claimed the 11th to the 13th century Crusades were unholy wars against a peaceful nation rather than holy wars, we had better take another look at these wars to retake Europe before consensus historians rewrite all of the eleventh century to excuse the villainous manner of the Muslim faith. I understand what they want to do. They are concerned that the "good Muslims" will be hated because "Muslim zealots" are pushing the Koran doctrine as Mohammed indicated in his cult documents and showing "Christian slaughtering hordes" seems to excuse the violence. The problem is that the premise they use is a lie hidden by another lie of consensus.

While our schools simply call the Wars "Holy Crusades", in the Age of Restoration, the Crusades were not only considered holy wars but also the retaking of Europe from the horrendous treatments of the Muslim overlords. It is true that the Muslim histories tell about how all the Europeans and Middle Easterners <u>welcomed the Muslims hordes with welcome arms and a smooth, bloodless transition occurred as Muslim Imperialism swept the countries of Syria, Jerusalem, Turkey, Greece, Italy, Spain, Sardinia, Turkey, the northern nations of Africa, and areas into regions of Pakistan and Iran</u>.

This is not true as the cut-off heads of those they defeated were paraded around to strike fear in the hearts of those who didn't want to worship a meteorite.

The attempted world takeover can be addressed as 9 holy wars or Jihads. What we call Crusades was simply a small portion of these Jihads where the Europeans were fighting back in a more unified manor.

622-634- Arabian Jihad-Mohammed's reign of terror included destruction of those in the Sinai Peninsula as he forced conversion to his new religion. Many died or were enslaved.

634-651-Persian Jihad- This bloody "holy War" included the takeover of Iran, Pakistan, and beyond.

634-1453-Byzantine Jihad- This Jewish and Christian focused Jihad included takeovers of Jerusalem, Syria, and Constantinople. This lasted for 800 years of almost continuous fighting.

636-1018-Indian Jihad – This was a bloody jihad to rid the world of Hindu and Buddhists.

640-700-North African Jihad- This included Tunisia, Libya and Egypt

651-751- Fertile Crescent Jihad- This included Iraq and Turkey which showed some of the bloodiest wars.

711-900- First Spanish Jihad- The takeover of Spain and Portugal; this was followed by years of killing to hold onto the countries.

720-732 French Jihad- This effort finally forced massive defeats and woke up the northern Europeans of the devastation that was about to come about.

812-940- Italian Jihad- This included the takeover of Corsica, Sicily, and Italy up to the Roman Catholic City. As the Catholic center was now in jeopardy, a union between Catholics and Byzantine was reached to get their counties back.

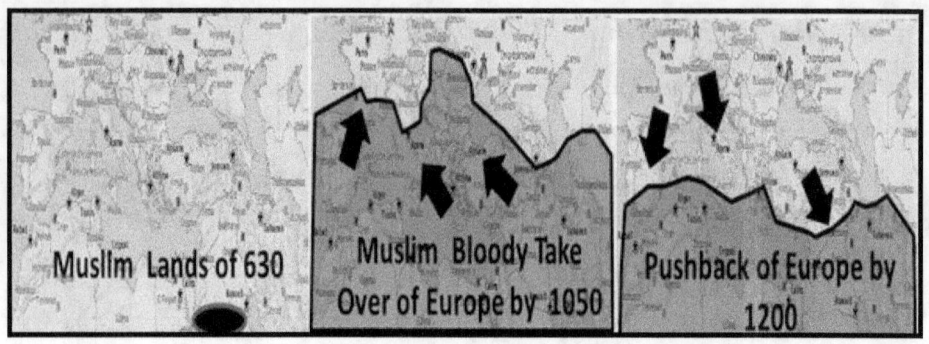

Generally speaking, non-Muslims were forbidden to write down the horrors, but one important manuscript did help us understand how they peacefully took over the land. Here is an excerpt of the "Chronicle of 754".

"Chronicle of 754"-Musa himself entered and long plundered Spain to destroy it. He built scaffolds and <u>decapitated</u> the Lords. The Spaniards only saw the <u>exposed sword</u>, <u>famine</u>, and <u>captivity</u>. He <u>burned</u> beautiful cities and condemned the powerful men to the <u>cross</u> and <u>butchered youths and infants</u> with the sword. Many flew to the mountains to risk hunger and various forms of death rather than give in. Abd al-Aziz took control of Seville and <u>made the queen marry him</u> and the <u>daughters of the kings and princes became his concubines</u>.

Please notice that what we call "the Crusades" actually the later part of the Muslim attempt to take over the world. Here are the more well-known ones taught in schools.

1095-98 First "State Religion" Crusade: *on command of Pope Urban II, thousands of Jews and Muslims were slain in Muslim held Hungary. While this was a horrible war, the Muslim invasion into Hungary had been halted for a time.*

1098-99 Second "State Religion" Crusade: *Areas of Muslim held Turkey were retaken from the Muslims and many* thousands were killed. By 1099, Muslim held Jerusalem was re-conquered and multiple thousands of Muslims and Jews were piled on the streets. The Muslim push into Europe was pushed back farther.

1204 Fourth "State Religion" Crusade: Moslem held Constantinople was retaken but at a horrible cost as thousands; from both sides lost their lives.

1212 The Children's Crusade- Just plain stupid as thousands of children march on the holy land only to be slaughtered and sold into slavery by the Muslims.

1229-1234 Crusades lost their unity and turned on themselves. It has been estimated that hundreds of thousands of Christian Cathar victims and the peasants of Germany were killed in very brutal ways for stating that what the New Testament called the Prince of the Earth could be considered an evil God. Other heresies of the State Religion included the Waldensians, Paulikians, Runcarians, Josephites, and many others. Most of these sects were exterminated; at least hundred thousand victims.

1260-1289-Ghingis Khan Crusades- Muslim held Palestine fought and managed to halt the advance of the Polytheist Mongols led by Genghis Khan and his descendants. They seemed to be a European ally in fighting the Muslims. By 1268 the Muslim had destroyed Antioch. Finally, the Muslims defeated the Mongols in 1281, but the Muslims were not finished, they turned to Tripoli and took it by 1289.

1290-In what was considered the last Crusade, a fleet of warships from Venice and Aragon arrived to defend what remained of the Crusader states in 1290 as Muslims tried to retake what the Crusades had freed.

1291- *Al-Ashraf, marched with a huge Muslim army against the coastal port of Acre, the effective capital of the Crusaders in the region since the end of the Third Crusade. After only seven weeks under siege, Acre fell, effectively ending the Crusades in the Holy Land after nearly two centuries and the Muslims reentered the lands.*

Because of the battles called the "Crusades" the onslaught of Muslims into Europe was reversed and inn some area they never regained control. That being said, the Muslim Imperialists were not finished.

Jihad After the Age of Crusades-While these will be the main Crusades discussed in this book, we can certainly see that the Muslim Crusades never really ended. Their bloodlust is remarkable and their devotion to Jihad is so very sad.

Ottoman Expansion Jihad *[1355-1690]* *Ottomans take control of the Muslim nations and decide to retake Europe, finally taking all of Turkey, Armenia, Greece and the Balkin regions of Europe and parts of Russia. Pushed back by "the Holy League [another Crusade]".*

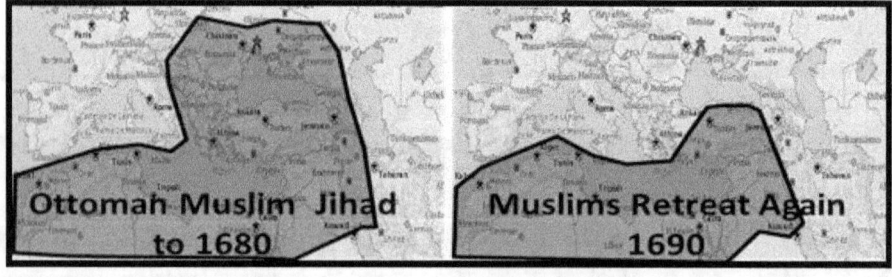

Barbary Pirate Jihad *[1801-1815]* *Muslims attack American, British, Swedish, and southern European ships for slaves, power and money. Our Marines still remember the shores of Tripoli and made it part of their fight song.*

Armenian Jihad [1895–1896]- *Over 200 thousand killed in mass executions to eliminate Armenian Christians from Turkey.*

Armenian Genocide and Jihad [1915-1919]- *In this fairly recent the Turkish Muslims massacred over* <u>***2 million Armenians by crucifixion, mass murder, and starvation***</u> *to try to eliminate the Christians. Starting in 1909, the Muslims were driven from the Balkans and were accepted by the Armenians as refugees.*

Greek Jihad [1919-1922]- *Muslim Turkey has expelled approximately 1 ½ million Greek Christians from its empire in the east and replaced them with Turks.*

Jew Elimination Jihad [1947-now]- *A half-dozen Arab countries have fought four wars in an attempt to destroy Israel and occupy its territory, and is currently continuing the attempt this very day with the publicly voted consent of 55 of the world's 57 Islamic nations. Muslims expelled approximately* <u>*800,000 Jews*</u> *from their homelands between 1947 and 1955 alone.*

Sudan Jihad [1956-1985]- *Muslim northern Sudan has conquered much of southern Sudan, killing and enslaving its Christian and pagan population.*

Cyprus Jihad [1974-1975]- *Muslim Turkey has invaded and occupied Northern Cyprus, over 200 thousand killed or displaced Greek Christians were removed.*

Timor Jihad [1975-1978]- *Muslim Indonesia invaded and conquered Christian East Timor and reduced the entire population by about 1/6. This very day, Muslim Indonesia is attempting to destroy Christianity in what used to be called the Celebes.*

Iranian Jihad [1980-1988] - *Muslim Iraq, in an imperialist act of aggression, invaded Muslim Iran with a resulting in the death of* **2 million people**.

Kurd Jihad [1987-2001] - *Tens of Thousands of Kurds killed by chemical and mass shootings to purify Muslims in Iraq and Turkey. The thousands and thousands of Kurds surviving Chemical warfare were permanently maimed.*

Hezbollah Jihad [1982-now] - *Hundreds are dead by children bombs, mortars, rockets, and terrorist action against Jewish people living in their own country. Muslim Government in Lebanon pays families of children who kill Jews by blowing themselves up.*

Hamas Jihad [1988-now] - *Hundreds are dead by children bombs, mortars, rockets, and terrorist action against Jewish people living in their own country. Muslim Government in Palestine pays families of children who kill Jews by blowing themselves up.*

Al-Qaeda Jihad [1988- now] - *Thousands including over 2000 Americans killed including blowing themselves up to kill Christians and Jews.*

Macedonian Jihad [2000-now] - *Muslim Albania, this very minute, is attempting to enlarge its borders at Christian Macedonia's expense.*

Muslim Brotherhood Jihad [2010-now] - *Thousands killed to reduce the Christian influence in Egypt, the Middle East and Northern Africa.*

Boko-Haram Jihad [2012-now] - *Thousands lay death and thousands more enslaved to force Muslim ideology into Nigeria and Chad.*

ISIS Jihad [2013-now]- *Thousands beheaded, burned alive, vaginas ripped open, Eyes popped out, enslaved, raped and killed to eliminate Christians from the world. To purify the religion, Chemical weapons are being used to kill Kurdish Muslims.*

We know the New Testament says "Turn the other Cheek" "Love and forgive your enemies", "Give unto Caesar what is Caesar's", and on and on, but what does the Muslim Bible tell them they MUST do? [They actually have 3 books in their Bible equivalent Quran, Hadith, and Sira]. Their "Sharia law is mostly established outside of the Quran and it is horrible to non-moon worshippers.

Hadith 553 *"the Apostle of Allah said,* **'Kill** *any Jew that falls into your power."*

Quran 8:12- *I will* **cast dread** *into the hearts of the unbelievers [Christians, Jews, Hindu, and Buddhist].* **Strike off their heads. Then strike off their fingertips**

Quran 8:16- *Muslim men who* **kill** *kafirs [Muslims who become Christians] avoid hell.*

Quran 9:5- *"When the sacred months have passed,* **slay** *the idolaters [Hindu] wherever ye find them, and take them captive, and besiege them, and prepare for them each ambush.*

Quran 9:29 **Attack** *the People of who read the Bible until they pay the jizyah, the dhimmi tax, submit to Sharia law and be humbled.*

Quran 9:39*: those Muslim men who refuse to fight and* **kill** *kafirs will be sent to hell.*

Quran 22:19-22- *"for the unbelievers, garments of fire shall be cut and there shall be poured over their heads boiling* water whereby whatever is in their bowels and skin shall be dissolved and they will be punished with hooked iron rods."

Quran 47:4- *"Therefore, when ye meet the Unbelievers in fight, **smite at their necks**; At length, when ye have thoroughly subdued them, bind a bond firmly:*

Quran-33:26- *"And Mohammed, the epitome of god, brought those of the People of the Book [readers of the Bible] who supported them from their fortresses and **cast terror** into their hearts, some of them you **beheaded** and some you **took captive**"*

Quran 48:29 *"Those who follow Muhammad shall **be ruthless** to the unbelievers but merciful to one another."*

Sharia Law - *"There is **no punishment for killing an apostate** since it is killing someone who **deserves to die**."*

Sharia Law - *"Jihad is a communal obligation upon Muslims each year. The caliph makes **war upon Jews, Christians, and Zoroastrians** until they become Muslim.*

Sharia Law- *Criticizing anything in the Quran or about Muhammad is apostasy and therefore **punishable by death**.*

Sharia Law: *-Any Muslim who states a preference for democracy rather than shariah law is a Kafir and therefore **sentenced to death**.*

Between 1990 and Today- The number of Special Islamic schools in America have quadrupled and the number of Islamic Mosques has tripled.

While you may not believe this as no news agencies are reporting the horrible actions of the bloody Muslims during this time. With chopped off heads, rapes, forced marriages, brutal killing, horrible taxes on those who they conquered, inability for those conquered to own business or raise their head when a Muslim passed, and destruction of all non-

Muslim religious buildings and artifacts, they "PEACEFULLY" took over southern Europe, Northern Africa, and much of the Middle East. They tried a second time in 1600 and continue today to fulfill what is commanded in their Bible. All those believing in their Koran are trying again. They have to and just because a few Muslims have rejected the teaching of Mohammed does not mean many will. If they wanted to, they would not be Muslims.

Conclusions

I hope your eyes have been opened a little bit and you are beginning to see how easy it is for a small group of scientists in a field can develop his own description of reality and many don't even question him even when his ideas are completely absurd. We need to open out eyes and try to retake sanity. There is a reason our constitution defines certain INALIENABLE- [or God-given] rights. That is because there is a Creator that must be above a government or the government WILL fall and the people will too easily be led like sheep to the slaughter by the powerful who are not restrained by any God.

I am old and an electro-optics Engineer by training, but that does not mean that I cannot see what is going on around me and you. We are being lied to. I know some lie to us to protect us, but most, simply lie to gain power over us or money from us. I don't like it. I especially hate how our children are getting more and more indoctrinated in fake consensus science. Soon, they will not be able to think. Every time they do, as most of the bunk has almost no theoretical backing, they are put down and told to do what is in the lesson plan. I found out first hand helping my grandson with his homework. I had not been indoctrinated by the same stuff when I was in school so he had to do some of it over. While I am seriously worried, I'm not sure how all of this is being taught today as I haven't been in school for a very long time, but I think what is done to hide every little thing that doesn't match up with the assigned science and history texts. It is really up to us to read what our children are being forced to learn and point out how horrible the texts are so that they can be changed. While I tried to steer away from religion and the huge secular path our schools are pushing, but while you are at it, see how lessons fit with your particular religion. Remember we do not have freedom from religion in our country, but instead, freedom to have religion.

Hopefully, you at least have questions with possibly solutions concerning the following:

1. The Earth is not as old as you were told ad nuclear decay timing has been debunked.
2. The probability that an undirected evolution process made humans has been debunked as the earth hasn't been around long enough to support it. Some type of "Directed" evolution appears to be the only reasonable possibility.
3. Civilized Giants once ruled the earth and were responsible to many horrible wars. Many were convinced these men were gods and that made a mess.
4. DNA mutation does not support "Out of Africa" statements but do support the possibility of the Bharata War and a massive war before the end of the Pleistocene.
5. Evidence suggests the art of flight is an ancient capability we lost in 3100BC when almost all major mutations of man that did not occur during the Pleistocene War happened according to Haplograph-DNA experts.
6. Apparently, dinosaur DNA was used to recreate the massive animals during the Pleistocene and into the Holocene.
7. People possibly, lived on Venus not too long ago.

8. Global warming has nothing to do with CO_2, but there is money telling everyone it does.
9. Edison was not as great an inventor as we were told in school.
10. We should not identify people by shape of the skull, bumps on the head, moles, slant of the nose, or distance between the eyes.
11. Just because a consensus of scientists tells us something, does not mean it is correct, especially, if the scientists may lose power, prestige, his job, or favor by going against others making a claim.

I certainly hope you enjoyed the book and with that I must end.

About the Author

Steve Preston is a long lime author of scientific, esoteric facts. His books focus on the painful truths rather than whitewashed details that make us comfortable. If you are interested in the truth instead of comfort, please review other works by Mr. Preston as shown next. The images are some of the author in Egypt taking the older version of taxi. To the right the writer is shown in the Jewish Negev desert of Israel where the Dead Sea Scrolls were found.

To the left below are a couple of pictures as we searched the New Zealand caves possibly visited by the ancient Maori and the last image is of the author investigating the statues on the Acropolis in Athens Greece.

Here is a partial list of some books that might be of interest.

Development of Mankind Series
The First Creation of Man-book 1 History of mankind
The Second Creation of Man-book 2 History of mankind
The Creation of Adam and Eve-book 3 History of mankind
The Antediluvian War Years-book 4 History of mankind
Man After The Flood-book 5 History of mankind
A Closer Look at Ancient History-book 6 History of mankind
A New View of Modern History-book 7 History of mankind
The Twentieth Century and Beyond- Book 8 History of Mankind

Bible History, Correction, and Analysis
Adam's First Wife-Story of Lilith
Expanded Genesis- Apocrypha and other Jewish texts
History Confirmed By The Bible- Science confirms the Bible
Moses Saved Egypt- How the Jews eliminated the Hyksos
Mysteries of the Exodus- Proofs of the Exodus
Old Testament Used By Jesus- Ancient Jewish texts
Understanding the New Testament-4th part of the Bible Series
Why the King James Bible Failed- Issues with KJB

Ancient Technology and Life
Ancient History of Flying- Ancient flying
Kingdoms Before the Flood- Pleistocene humans
Living on Venus- Venus before the Pleistocene Extinction
Martians- Ancient Life on Mars
Mysterious Pyramids- Who made the Pyramids?
Not from Space- UFOs are not from space.
Amazing Technology- Descriptions of prehistoric capabilities

Ancient and Modern War
America's Civil War Lie- Truth about the Civil War years
Behind the Tower of Babel- Story of the Bharata War
Driven Underground- Fear in the Bharata War
World War Before- The Pleistocene War
World War with Heaven- The Angel and Anak War
World War Zero-The Bharata War
When Giants Ruled the Earth- History of the Titan Giants

Current Events and Fears
Allah' God of the Moon- Terror of Muslims
American School Disaster- fear in our country
Scythians Conquer Ireland- A History of Ireland

Make Your Own Global Warming
Truth About Phoenicia- The Evidence -First in America
Our Very Odd Presidents- 60+ President reviews
Terror of Global Warming- Fake issue is uncovered
Humans on Display- Slavery and Human Zoos

New Look at Physics
Amazing Technology- Pleistocene Technology
Fake Science- Use of Science to control People
Is Time Travel Possible? Science of Time Travel
Releasing Your Consciousness- Beyond our SELF
Slip Through a Wall- How-to walk-through solids
Our 12-Dimensional Universe- New science of our Universe
Mystery of Photons and Light- Science of Photons
Vibrational Matter- New Science of Quantum Fluctuations

New Look at Biology
DNA of Our Ancestors- Tracing DNA of ancient man
God Didn't Make The Ape- New science on ape Evolution
Lizard People- Mutated People of the Bharata War
Races of Men- Tracing DNA of Humans
Self-Virtualization- New science of reality
Life Resonance- Unusual capabilities of men
Biophotonics and Healing- How Photonics used in medicine

www.ingramcontent.com/pod-product-compliance
Lightning Source LLC
Chambersburg PA
CBHW072028230526
45466CB00020B/1091